ACS SYMPOSIUM SERIES 436

The Beilstein Online Database

Implementation, Content, and Retrieval

Stephen R. Heller, EDITOR
U.S. Department of Agriculture

Developed from a symposium sponsored
by the Division of Computers in Chemistry
at the 198th National Meeting
of the American Chemical Society,
Miami Beach, Florida
September 10-15, 1989

American Chemical Society, Washington, DC 1990

Library of Congress Cataloging-in-Publication Data

The Beilstein online database: implementation, content, and retrieval /Stephen R. Heller, editor

p. cm.—(ACS Symposium Series, 0097–6156; 436).

"Developed from a symposium sponsored by the Division of Computers in Chemistry at the 198th National Meeting of the American Chemical Society, Miami Beach, Florida, September 10–15, 1989."

Includes bibliogralphical references and index.

ISBN 0–8412–1862–5

1. Beilstein, Friedrich Konrad, 1838–1906. Beilsteins Handbuch der organischen Chemie—Congresses. 2. Chemistry, Organic—Data bases—Congresses. 3. Information storage and retrieval systems—Chemistry—Congresses. I. Heller, Stephen R. 1943– . II. American Chemical Society. Division of Computers in Chemistry. III. American Chemical Society. Meeting (198th: 1989: Miami Beach, Fla.). IV. Series

QD257.7.B39 1990
547′.00285′574—dc20 90–41494
CIP

PRINTED IN THE UNITED STATES OF AMERICA

ACS Symposium Series

M. Joan Comstock, *Series Editor*

Foreword

THE ACS SYMPOSIUM SERIES was founded in 1974 to provide a medium for publishing symposia quickly in book form. The format of the Series parallels that of the continuing ADVANCES IN CHEMISTRY SERIES except that, in order to save time, the papers are not typeset, but are reproduced as they are submitted by the authors in camera-ready form. Papers are reviewed under the supervision of the editors with the assistance of the Advisory Board and are selected to maintain the integrity of the symposia. Both reviews and reports of research are acceptable, because symposia may embrace both types of presentation. However, verbatim reproductions of previously published papers are not accepted.

Contents

Preface vii

1. **The Beilstein Online Database: An Introduction** 1
   Stephen R. Heller

2. **Computerizing Beilstein** 10
   Clemens Jochum

3. **STN Implementation of Factual and Structure Databases** 24
   Andreas Barth

4. **An Overview of DIALOG** 42
   Ieva O. Hartwell and Katharine A. Haglund

5. **Chemical Structure Searching: Using S4/MOLKICK on DIALOG** 64
   Stephen M. Welford

6. **Support of Industry Information Needs** 80
   Norman J. Santora

7. **Searching for Chemical Reaction Information** 88
   Damon D. Ridley

8. **Physical Property Data: Capabilities for Search and Retrieval** 113
   Andreas Barth

9. **Experiences of Two Academic Users: Evaluation of Applications in an Academic Environment** 130
   Gayle S. Baker and David C. Baker

10. The Lawson Similarity Number (LN): Offline Generation and Online Use 143
Alexander J. Lawson

Author Index 157

Affiliation Index 157

Subject Index 157

Preface

WHEN THE ACS DIVISION OF COMPUTERS AND CHEMISTRY decided to hold a symposium on numeric databases in chemistry, several possible speakers were considered, including Clemens Jochum, President of the Computer Division (COMP) of the Beilstein Institute. The organizers quickly concluded that the size and impact of other databases under consideration, as well as attempts to squeeze all the activities of the Beilstein computerization efforts into Dr. Jochum's talk, combined to create an impossible task. Therefore, with the assistance of Clemens Jochum and Sandy Lawson of the Institute, and Bob Badger and Arnoud DeKemp of Springer–Verlag, the organizers decided to focus upon the number of areas of numeric, factual, and structural information in the massive Beilstein computerization effort.

All the speakers from the symposium agreed to contribute to this book, and additional chapters were added to provide coverage of areas not included therein. Andreas Barth was persuaded to write two separate chapters in order to make this book more useful to both the casual user of the STN system and the expert interested in the physical property data in Beilstein Online. The result is a book that, for the first time, provides an overall picture of the entire Beilstein computerized operation and the results now available both to organic, analytical, and physical chemists, and to chemical information specialists.

The Beilstein Online Database covers implementation of the database, its content, and how it can be searched on two major online systems from the perspective of industry and academia.

Years ago I wrote that Beilstein Online is a renaissance or reawakening of a sleeping giant. I believe the contributors to the book have proven this to be true. I hope readers will again make extensive use of this unique resource in their organic chemistry research activities.

STEPHEN R. HELLER
U.S. Department of Agriculture
Agricultural Research Service
Beltsville, MD 20705–2350
June 11, 1990

Chapter 1

The Beilstein Online Database

An Introduction

Stephen R. Heller

U.S. Department of Agriculture, Agricultural Research Service, Northeastern Region, Beltsville, MD 20705

One of the most exciting events in chemical information has been the conversion of the *Beilstein Handbook of Organic Chemistry* (*1*), commonly referred to as *Beilstein,* into computer-readable form and its availability as an online database. This event is quite recent and has generated considerable interest in the chemical community. Because no broad-based description of the online Beilstein database exists, a book on this subject will be of value to various groups in the chemical community, including organic chemists, information specialists, and those in the académe. This book, based on a symposium (*2*), is the first collection of papers discussing the important aspects of the computer-based version of the *Beilstein Handbook.*

The *Beilstein Handbook,* the most complete and systematic collection of evaluated data on organic compounds, consists of over 350 printed volumes comprising more than 275,000 pages of text. A chemical is included in *Beilstein* if it satisfies the following three requirements. First, the chemical must be an organic compound. Second, it must have a known, verified structure and must be able to exist as a pure compound. Third, a fully described method of preparation and some published physical or chemical data about the compound must be available. Entries in the *Beilstein Handbook* come from the chemical literature including journals, from patents, and from monographs.

To appreciate the enormous task involved in making the Beilstein database available online, a brief history is in order. *Beilstein,* or more accurately, the standard reference work known today as *Beilsteins Handbuch der Organischen Chemie,* is a descendant of the original *Handbuch*, whose first edition was created by Friedrich Konrad Beilstein in St. Petersburg in 1881.

Professor Beilstein, born in St. Petersburg, Russia, of German parents in 1838, was educated at the Universities of Heidelberg, Munich, and Goettingen and assumed a professorship at the Imperial Technical Institute in St. Petersburg in 1866. The first edition of *Beilsteins Handbuch der Organischen Chemie* was published in 1881–1882 and consisted of two volumes for a total of 2200 pages, in which about 15,000 organic compounds were described. Beilstein published a second edition (three volumes, 4080 pages) between 1885 and 1889 and a third edition (eight volumes, 11,000 pages) between 1892 and 1906, the year of his death.

The size and scope of the *Handbook* was such that it could no longer be managed by one individual, and accordingly, the German Chemical Society undertook this responsibility after Beilstein's death. The publication of the current edition of Beilstein, the fourth edition, began in 1918, under the editorship of P. Jacobson and B. Prager. In 1933, F. Richter was named as editor, and he was followed in 1961 by H.-G. Boit. The current editor, R. Luckenbach, succeeded Professor Boit in 1978. The *Beilstein Handbook* is distributed by Springer–Verlag publishers (*3*).

The fourth edition of *Beilstein* is the basis of the Beilstein Online database. It consists of a main work (*Hauptwerk*) and five supplementary series (*Ergaenzungswerke*). The combined supplement E III/IV is not regarded as a separate supplement. Each of these supplementary series covers different time periods, as shown in Table I. The basic work and the first four supplements were published in German, but the fifth supplement is in English, as will be all future supplements.

Table I. Organization of the *Beilstein Handbook*, Fourth Edition

Series	Period of Literature Covered	Abbreviation
Basic Series	1830–1909	H
Supplement I	1910–1919	E I
Supplement II	1920–1929	E II
Supplement III	1930–1949	E III
Supplement III, IV	1930–1959	E III/IV[1]
Supplement IV	1950–1959	E IV
Supplement V	1960–1979	E V

[1] Volumes 17–27 of the Supplemental Series III and IV, covering the heterocyclic compounds, were combined into a joint issue as a result of some disruptions occurring in Germany during the period involved.

The main work and each of the supplementary series consist of 27 "volumes," each of which may be one or more physical books. Which compounds appear in which volume is determined by the compound's chemical structure, as described in the Beilstein classification system procedures. Table II shows the main divisions of the *Beilstein Handbook*.

A central feature of the *Handbook*, therefore, is the way in which it is organized. A specific structure will be found at essentially the same place in any of the supplementary series. In the past, determining that location

Table II. Main Divisions of the *Beilstein Handbook*

Compound Group	*Volume Numbers*
1. Acyclic compounds	1–4
2. Isocyclic or carbocyclic compounds	5–16
3. Heterocyclic compounds	17–27

from the structure required a knowledge of the Beilstein systematic rules for filing. These rules are well described in a brochure published by the Beilstein Institute and entitled "How to Use Beilstein," which is available from either the Beilstein Institute or Springer–Verlag publishers (*1, 3*).

A given compound will appear in the same volume in each series; thus thiophene is found in Volume 17 of the main work and Volume 17 of each of the supplementary series, because Volume 17 is devoted to heterocyclics containing one chalcogen (Group VI; O, S, Se, or Te) heteroatom.

The main consequence of this organization is that the *Beilstein Handbook*, rather than being a linear chronological record, is really a series of "snap-shots," taken at 10- or 20-year intervals, of the entire organic chemical world.

Today, one can use the computer program SANDRA (Structure and Reference Analyzer) (*3*) to locate a system number when looking up a compound in the *Handbook*. Alternatively, the Lawson Number (*see* Chapter 10) can be used with the online system. When a chemical is retrieved from Beilstein Online, the *Beilstein* citations—typically one per series—are provided. These citations are in the form 4–17–00–00093, which indicates page 93 of Volume 17 of the fourth supplementary series.

The heart of the *Beilstein Handbook* is the factual information associated with each compound. Each published series contains new information about a compound, which means that a searcher may have to look into a number of volumes to find all the information on a given compound. In contrast, in the online database, data from the main work and all the supplementary series are combined to form a record that encompasses all the available knowledge about structure. The online version includes both the complete full records (Full File), which contain evaluated data, and the partial incomplete records (Short File), which contain unevaluated information. The incomplete records provide a more up-to-date database for searching. Each full record of a compound in the Beilstein database contains the following information:

- Identification, structure, and configuration
- Natural occurrence, and isolation from natural products
- Preparation and purification
- Physical properties
- Chemical properties
- Analytical and characterization data
- Related salts and addition compounds

The total number of types of data that *may be* available for a compound is in excess of 200, such as for pyridine. Each of these fields may be searched and displayed. However, very few compounds have many fields of information actually in the record associated with the chemical. The exact details of searching and retrieval vary among the different online host implementations of the Beilstein database. At present *Beilstein* is available online on STN (*4, 5*) and DIALOG (*6*), and by mid-1990, it will be available through the Maxwell Online–ORBIT system (*7*).

In addition to all the factual data, which are extensively reviewed and cross-checked before they are added to the database, a record is kept of the source of the data. A literature citation for every measurement of a datum is provided. It is possible also to search through these literature citations, but only on the basis of the first author's surname.

Implementation of the Database

In contrast to the weekly updates of the CAS (*Chemical Abstracts Service*) Registry file, the Beilstein Structure and Factual Database files are currently being updated on a much less frequent basis. In addition, the entire *Beilstein Handbook* is not yet available online. The schedule for implementation of the computer-readable form of the database is indicated in Table III.

TABLE III. Implementation of Beilstein Online Database

Step	*Year Available*	*Class of Compounds*
1	1988	Heterocyclic (Volumes 17–27) (Full File)
2	1989	Heterocyclic (Volumes 17–27) (Short File or excerpts)
3	1989	Acyclic (Volumes 1–4) (Full File)
4	1990	Isocyclic (Volumes 5–16) (Short File or excerpts)
5	1991	Isocyclic (Volumes 5–16) (Full File)

Online Access to Beilstein

As indicated in Table III, the first part of the database of the file to be mounted and made searchable online was the full portion on heterocyclic compounds of the *Beilstein Handbook,* Volumes 17–27. The period of the literature covered was from 1830 to 1959. Since then, the so-called Short File, or excerpts, of additional information and additional heterocyclic compounds has been made available online. The Short File, or excerpts, includes compounds from the scientific literature from 1960 to 1979 that have yet to be critically reviewed and evaluated. These compounds correspond to the organic chemicals that are found in the printed *Beilstein Handbook,* Supplementary Series V. The initial Full File of heterocyclic compounds compiled from 1830 to 1959 encompassed about 350,000 compounds. The Short File compiled from 1960 to 1979 added an additional 2.65 million compounds to the database, to bring the total to a little over 3 million compounds. At this time, neither the Full File or Short File contains any salts of the compounds.

Since the end of 1988, the first part of the Beilstein database has been available online through the STN–International system. The actual STN host computer on which the database is mounted is in Karlsruhe, the Federal Republic of Germany (*4*). At the end of 1989, the Beilstein database became available online on the DIALOG system. By mid-1990 it will also be available through the Maxwell Online–ORBIT system. All three systems are provided with the same database and chemical structures on computer tapes by the Beilstein Institute in Frankfurt/Main. Details of the implementations of the STN and DIALOG versions of Beilstein Online can be found in Chapters 3 and 4, respectively. The following sections discuss some overall comments about the three systems.

Each of the online vendors has different data and text search software. To handle the types of numeric and factual data found in the Beilstein database, all three vendors had to develop additional search software capabilities. As with bibliographic searching, the differences in these three systems are both objective and subjective. It is not the purpose of this chapter to evaluate the search software of these three vendors. Indeed it would be of little value, because searchers tend to have their own criteria or requirements for searching a particular vendor or vendors. However, an overall view of the systems available on each vendor will be described so that should a particular feature be of importance to a user, it will be properly explained here.

The main differences between the three vendors is the kind of structure-searching and display software used for exact and substructure searching. The three systems are summarized in Table IV.

Table IV. Chemical Structure Searching Software for Beilstein Online

Vendor	*Software*	*First Commercial Use*	*Reference*
STN	CAS ONLINE	1981	5
DIALOG	S4	1989	8
ORBIT	HTSS	1987	9

It is interesting that none of the vendors chose the DARC software, available on the Questel online system, which was one of the earliest (first used commercially in 1980) structure-searching software systems to be available commercially. The NIH/EPA CIS SANSS (10) software (first made commercially available in 1977) was also not chosen by any vendor for this database.

From the viewpoint of the end user or the online searcher, the variety of software packages for structure searching offers a unique opportunity to examine the different algorithms and methods of structure searching, and it is possible that, for some types of searching, one system may prove superior over another. The availability of these three systems will undoubtedly provide a fertile ground for a number of interesting studies on the structure-searching capabilities of different software algorithms.

Costs

The cost of online searching has been rising over the years. From an hourly royalty of some $4 in the early 1970s to what is now well over $100, users have seen their online search system budgets increase at a rate much faster than that of inflation in most countries. The cost of searching bibliographic databases is now about $135 per hour, on the basis of 1989 prices from the three vendors who will be providing Beilstein Online. Structure searching of the CAS 9 million plus files of connection tables costs about $250–300 per hour.

The current cost of the Beilstein Online factual and structure files averages to $225–250 per hour for both STN and DIALOG, although the cost can go considerably higher, depending on the exact nature of what one does on the system. The system on STN has various connect time, search term, and other charges, so the average cost will depend on the exact type of search conducted. The DIALOG system charges a fixed hourly connect time price, and the cost of using it for data searching is thus independent of the type of search performed.

However, some additional fees for certain records and structures being printed out may be charged. ORBIT has chosen a third and different method of charging. However, as one vendor candidly pointed to a user grumbling about the cost of the Beilstein Online, all the vendors really charge essentially the same fees but use different algorithms. The reason for this is simple. The contracts each of the three vendors signed, which allow them to make the Beilstein database available, require a certain royalty revenue for usage. Thus, practically speaking, $250 per hour is the most likely cost you will incur on any of the three systems.

Although this amount may, at first glance, be regarded as high, the value for the money, or the cost-benefit analysis, is considered to be well worth the price, because in a bibliographic database, one must first perform a search and then look up the references or literature citations to find the actual data desired. The extra cost of using Beilstein Online pays for the "preprocessing" of each literature citation, which involved the intellectual extraction and evaluation of the data in the original article. Furthermore the data are then compared with other values found in other published articles for the same property for the chemical. This massive high-level, high-quality, and very labor intensive effort, which the Beilstein Institute must pay in salaries for its Ph.D. chemists, must be recovered in some way. Although the government of the Federal Republic of Germany has provided subsidies over the years, in the near future the Beilstein Online database must recover its obvious high costs from its only real source—the user community.

Outline of This Book

This book consists of 10 chapters. This first chapter is an introduction to and an overview of the Beilstein Online database and provides a short description of the computer systems and related search software that were operational as of the date of the symposium. The next chapter, by Clemens Jochum, the president of the Beilstein Institute computer division, details how the database was assembled, its current status, and the future of the computer activities at the Beilstein Institute.

Chapters 3 and 4 describe the specific implementations of the two currently operational systems, STN and DIALOG, respectively. The STN chapter, "The STN Implementation of the Beilstein Factual and Structure Databases," was written by Andreas Barth, the person responsible for the implementation of the system for STN. The chapter describes the types of information in the database on STN and gives numerous search examples. The DIALOG chapter, "Beilstein on DIALOG," written by Kathy Haglund, the person responsible for the implementation of the system for DIALOG, is written in a similar fashion, with examples of how one performs a search on DIALOG. This chapter explains how DIALOG has indexed the Beilstein database and how it can be searched. One nice feature of the DIALOG implementation is the use of the S search operator to do a combined search for a property with a related refinement, such as boiling point at a given pressure or heat of formation at a specific temperature. In addition, the DIALOG implementation describes how structures can be created, searched, and displayed by using a combination of the MOLKICK (*11*) program or the ROSDAL linear-notation structure strings, S4 structure searching, and the DIALOG GEOFF (Graphics Enhanced Output File Format) structure display program.

Besides the introduction in Chapter 4, little published information is available about the newly released S4 chemical structure search software system that DIALOG is using. This system is available for the first time to the online community. Chapter 5 is devoted to an extended discussion of the S4 system. Included in this chapter is a discussion of the MOLKICK (*11*) PC-based software package that allows one to easily draw structures for up-loading to the DIALOG mainframe computer for structure searching. The Wiswesser-type linear-notation scheme used by the MOLKICK program, the ROSDAL (Representation of Organic Structure Descriptions Arranged Linearly) string, which is how structures are entered as queries into the S4 search system, is briefly mentioned.

Chapter 6 gives an industrial view from a large pharmaceutical company on how they use the Beilstein Online database. The examples in this chapter relate to using the Beilstein Online database as an aid in drug design research. A novel application of the use of the Lawson Number (described in detail in Chapter 10) is also presented.

The next presentation, and by far the longest in this book, is by Damon Ridley. Chapter 7 describes the very valuable chemical reaction information that is part of the Beilstein Online database, with examples of how one can search for this information.

Chapter 8 examines another area of strength of the Beilstein Online database, the physical property data in the system. In his second chapter in this book, Andreas Barth gives a detailed account of data range searching for numeric data and what range searching really means as implemented on STN. A number of examples are given in a number of figures in the chapter, showing how a particular search either gives a hit or a miss. The more sophisticated data searches are explained. From this discussion, the reader will be able to understand better the differences between checked (evaluated) and unchecked data in the database. The very useful and important ability of the STN system to accept different units and to convert units within the system is also well described.

Chapter 9 is a view from academia relating the experience of an organic chemist in a chemistry department and of a librarian in a chemical engineering department, both of whom have been using Beilstein Online since it was first available.

Finally, Chapter 10 describes a valuable and new tool for structure similarity searching in Beilstein Online. The Lawson Number (LN), the computer system equivalent of the Beilstein classification system number, is valuable for many types of searches for particular types of organic chemical structures. This chapter very clearly explains the significance of the LN, why compounds have more than one LN, and how the new version of the SANDRA program (*3*) allows one to create the LNs for any structure drawn.

Literature Cited

1. Beilstein Institute, Varrentrappstrasse 40–42, D–6000 Frankfurt/M 90, Federal Republic of Germany. The Beilstein Help Desk phone number is 49–69–7917–258. The FAX number is 49–69–7917–492.
2. "Experiences with Beilstein Online," sponsored by the Division of Computers in Chemistry of the American Chemical Society, Fall ACS Meeting, Miami, FL, September 12, 1989.
3. a) Springer–Verlag New York, 175 Fifth Avenue, New York, NY 10010. The Beilstein phone number is (212) 460–1622. The FAX number is (212) 473–6272.

b) Springer–Verlag, Tiergartenstrasse 17, D–6900 Heidelberg, Federal Republic of Germany. The phone number is 49–6221–487–457. The FAX number is 49–6221–43982.

4. STN–International, Postfach 2465, D–7500 Karlsruhe 1, Federal Republic of Germany. The phone number is 49–7247–82–4566.
5. STN–International, CAS, PO Box 02228, Columbus, OH 43210. The phone number is (800) 848–6538 or (614) 447–3600. The FAX numbers are (614) 447–3709 or (614) 447–3713.
6. DIALOG Information Services, 3460 Hillview Avenue, Palo Alto, CA 94304. The phone number is (800) 334–2564 or (415) 858–2700.
7. Maxwell Online–ORBIT Search Service, 8000 Westpark drive, McLean, VA 22102. The phone number is (800) 456–7248 or (703) 442–0900. The FAX number is (703) 893–4632.
8. S4 is available from Softron GmbH, Rudolph Diesel Strasse 1, D–8032 Grafelfing, Federal Republic of Germany. The phone number is 49–89–855–056. The FAX number is 49–89–852–170.
9. HTSS is available from ORAC Ltd., 18 Blenheim Terrace, Woodhouse Lane, Leeds LS2 9HD, United Kingdom. The phone number is 44–532–441–821. The FAX number is 44–532–448–283.
10. G. W. A. Milne, S. R. Heller, A. E. Fein, E. F. Frees, R. G. Marquart, J. A. McGill, and J. A. Miller, "The NIH/EPA Structure and Nomenclature Search System (SANSS)," *J. Chem. Inf. Comput. Sci.,* **1978,** *18,* 181–185.
11. MOLKICK is available from Springer–Verlag. *See* references 3a and 3b.

RECEIVED February 10, 1990

Chapter 2

Computerizing Beilstein

Clemens Jochum

Beilstein Institute, Varrentrappstrasse 40–42, D-6000, Frankfurt am Main, 90, Federal Republic of Germany

The contents of the Beilstein Information System are described. The database structure is designed on the basis of the printed Beilstein products. After a detailed description of the database structure, the production process for the database is explained. The database can be searched by structure/substructure, numerical fields, keywords, or a combination of these search methods. Various access methods are available via public online hosts or through private in-house implementations.

Keywords: Data Structure, Factual Database, Beilstein, Organic Chemistry, Numeric Data, Connection Tables.

The Beilstein Database Project was started in October 1983 [1,2]. Its goal was to build the world's largest organic factual database. After 15 months of systems analysis the venture went into full production in January 1985. There are three main sources for the database:

- The Beilstein Handbook with more than 340 volumes containing more than 1 million title compounds and the associated factual data including preparations. This data source covers the primary literature time frame from 1830 to 1960.

- 6 Million Beilstein Filecards containing compounds and factual data of the primary literature time frame 1960 to 1980. This data source contains ca. 3 million additional compounds.

- Compounds on factual data of the primary literature starting from 1980 till now.

0097–6156/90/0436–0010$06.00/0

The input of the first two sources has been completed in 1988. The third source is currently being input. All input is carried out using personal computers. The input software has been developed at the Beilstein Institute. In addition software for loading the data and structures on the mainframe, registry software and retrieval software for structures and factual data has been developed.

All heterocyclic compounds from the first two sources covering the literature period from 1830 to 1980 are already available online. Online availability of the acyclic and isocyclic compounds of these two sources will be accomplished by the middle of next year. Subsequently structures and factual data of the primary literature from 1980 onwards will be made available on an ongoing basis.

1. The Sources of the Beilstein Information System.

The articles of the Beilstein Handbook, i.e. the complete factual descriptions of a compound, have always been written according to a very well defined structure. Naturally, since the analytical methods and chemical preparations have changed over the last fifty years, the instructions and definitions for the Beilstein manuscript writers have been altered slightly accordingly over this period of time. These changes had to be taken into consideration for the definition of a computer-optimized data structure.

After a very thorough analysis of the article structures of the main volume and of all supplementary series, a data structure was defined which allowed the computerized input, storage and retrieval of the Handbook data without loss of information. However, some compromises had to be made for most organic compounds described in the primary literature, only very little factual information is known. In many cases, only the boiling point, melting point, refractive index and one or two methods of preparation have been described in the literature. These small information compounds (SIC) usually require in their redundancy- and error-free Beilstein presentation only a quarter of a page or less in the Handbook without any loss of information from the primary literature.

A comparatively small percentage of all known compounds (less than 5%) are very important for chemical reactions or pharmaceutical purposes and were therefore published widely in the chemical primary literature. Therefore many physical data, preparations and other factual data are known for these large information compounds (LIC). The definition of the data structure had to take the different information contents of SICs and LICs (and all variations in between) into account. The final electronic data structure constitutes a very sophisticated compromise between the information contents of LICs and SICs:

- The database contains the complete Handbook information of the SICs and compounds with an intermediate amount of information.

- For the LICs, only a subset of all Handbook data can be stored in the electronic database. For more information, online users are then referred to the Handbook.

Since the information contents of the Handbook and the database are partially overlapping, Handbook subscribers can access the database at a reduced rate.

In addition to the Handbook, the database contains information from another source:

The literature of the Fifth Supplementary Handbook Series (literature time frame: 1960-1980) has been completely abstracted. This factual information which is contained on 7.5 million file cards (one card per compound per literature citation) is the basis for the Handbook articles of this series. Since this Handbook series will not be completed for a further decade, the online user will be provided with access to the "raw" information. In contrast to the Handbook data, this information contains redundancies (from different articles about the same compound), errors (from the primary publication, which could only be removed by crosschecking with other sources) and missing data (the file cards only contain the 5 most important physical data and the preparation).

The input of the Handbook and the Filecards have been completed. Most of this information is already available online or will be made available in the near future. (See Fig. 1)

The Beilstein Institute is currently working on its third source of information: The factual data of the primary literature from 1980 onwards. The compounds and associated literature data are abstracted in a completely new electronic and paperless way using off-line microcomputers. By using Excerp, a program especially developed for our needs, the abstractor graphically inputs the structure and via a menu-driven intervace also inputs the factual information. The design of the data structure took new developments of analytical and synthetic methods into account.

The structure of the database can be divided into two parts: The numeric factual file and the structure file. These two parts are subsequently described in more detail.

2. The Factual File.[4]

The Factual File has a relational structure and contains three types of fields:

1) **Numerical Fields.** Most numerical parameters can be stored as 2-byte integers, but some physical parameters require a 4-byte floating point format.

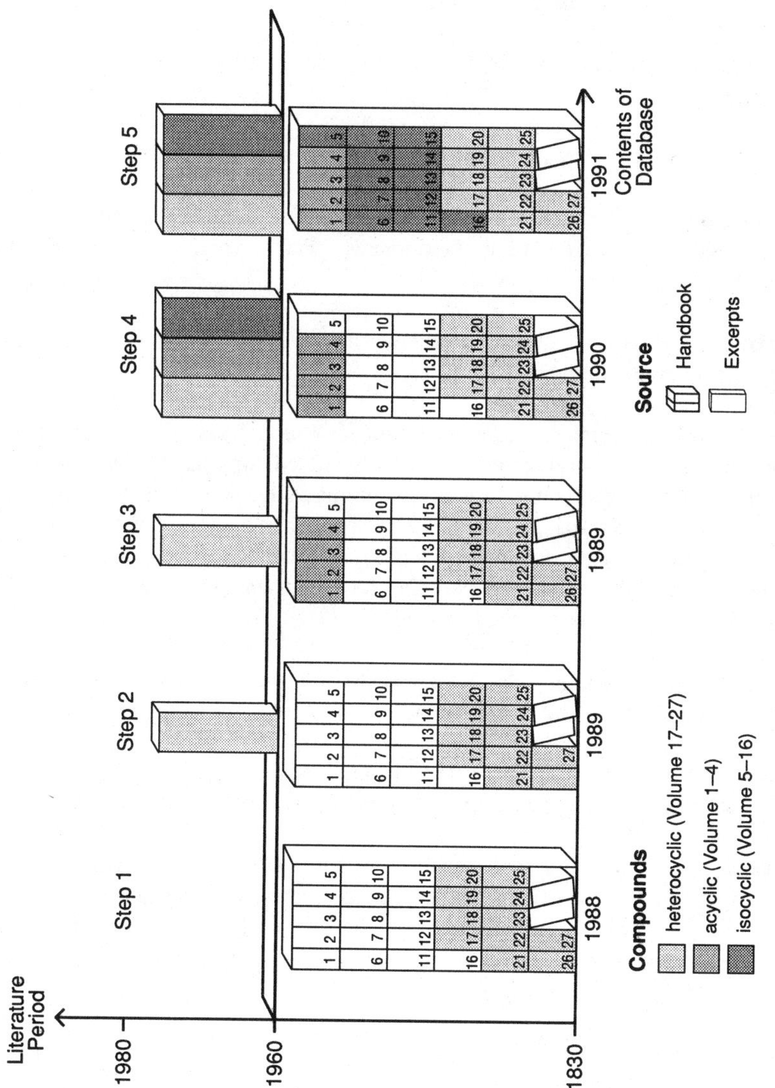

Figure 1. Status of input of the Handbook and Excerpts.

2) **Boolean Fields.** These fields store the presence (or absence) of a keyword or a parameter.

3) **Alphanumeric Fields.** Literature citations, chemical names, comments for a preparation, etc. are stored as character strings.

All boolean fields and most of the numerical fields are searchable separately or in combination.

Chemically, the data structure can be divided into 7 parts:

1) **Identifiers.** These parameters contain the molecular formula and registry numbers for identification and search of the compounds. Three registry numbers are stored for each compound:

- The Beilstein Registry Number is a structure-independent compound-identifier for the internal organization of the data base.

- The Lawson Number is a structure-dependent non-unique hash-code. Since this number is structure-related, several structurally closely related compounds can have the same Lawson Number. The structure-ordering according to this number is very similar to the Beilstein Handbook Ordering System. The number can be computed on a microcomputer (after having entered the structure graphically) using a rather complex algorithm. Subsequent searching of this number on an online host represents an elegant and inexpensive way of structure and substructure browsing.

- The CAS registry number. This number is also not structure-related but allows easy cross-file searching on other databanks.

2) **Structure-related Data.** These fields include information about the purity of a compound, its possible tautomers or alternative structure representations.

3) **Preparative Data.** This topic includes all preparation-related parameters such as starting material, yield, solvents, temperature, pressure, by-products, etc.

4) **Physical Properties.** This division includes most of the numerical fields. It is subdivided into:

- Structure and Energy Parameters (Dipole Moment, Molar Polarization, CouplingConstants, etc.),

- Physical Properties of the Pure Compound (Melting Point, Boiling Point, Transport Phenomena, Calorical Data, Optical Properties, Spectral Information, Magnetic Properties, Electrical Properties, Electrochemical Behaviour, Colour),

- Physical Properties of Multicomponent Systems (Solution Behaviour, Liquid/ Liquid-, Liquid/Solid-, Liquid/Vapour-Systems, etc.).

5) **Chemical Behaviour.** Reactions of this compound with other chemicals are described under this section. The fields include reaction partners, reagents and reaction conditions.

6) **Physiological Behaviour and Applications.** Use-, Toxicity-, Biological Function- and Ecological Data-Parameters are described under this topic.

7) **Characterization of Derivatives and Salts.**

There are 510 fields in total. 305 of these fields are searchable separately or in combination (208 boolean fields and 97 numerical fields).

3. The Structure File.[5]

The structures of the Beilstein-compounds are stored as connection tables (CTs) to allow a very flexible structure and substructure search. Since most commercially available structure/substructure handling programs such as MACCS (MDL) or DARC (Telesystems-Questel) work on the basis of CTs, the Beilstein Registry Connection Table (BRCT) can be easily adapted for Inhouse-Systems.

The following criteria governed the design of the BRCT format.

(i) Completeness and Unambiguity

The Registry CT must contain all of the structural information which is necessary to completely describe the chemical structure of a compound for the purpose of its storage and retrieval.

By virtue of its completeness, the Registry CT must provide an unambiguous representation of each compound in the Registry file, so that a Registry CT describes one and only one compound in the Registry file [3].

(ii) Uniqueness

The Registry CT must provide a unique representation of the structure of each compound in the Registry file, so that each compound has one and only one Registry CT.

(iii) Compactness

The Registry CT must be compact, in order to minimise the storage requirements of the

Registry file, and to minimise the quantity of data which may be required to be transferred between different applications programs.

(iv) Flexibility

The Registry CT must be flexible, although precisely defined, in order to enable upgrading or other modification of applications programs without requiring the Registry CT to be redefined.

(v) Treatment of tautomerism and resonance

The Registry CT must provide for a satisfactory treatment of the phenomena of tautomerism and resonance. So that different "valence bond" descriptions of the same molecule can be recognised as equivalent and can be represented in the Registry file by the same Registry CT and different tautomeric forms of a molecule to be represented by different Registry CTs in cases where this is desirable.

The Beilstein Registry CT consists of a Header vector and a number of "lists". The Header vector is obligatory, and must be present in the CT. Certain of the lists are also obligatory, while others are optional and are present in the CT only when necessary.

The Header vector contains information which controls the length of the obligatory lists, and the number of optional lists which are present in the CT. The lists themselves describe the structure graph, modifications of the structure graph, and any supplementary information which is necessary to completely describe the structure of the chemical compound.

The Header vector is of a fixed length, which is defined by the BRCT format. The lists are of variable length, with the result that the BRCT itself is of a variable length. The length of each obligatory list is known from the Header vector, and need not be stored explicitly. The length of each optional list must be stored explicitly with each list in the CT.

The following lists are used in BRCT Version 1.00:

3.1 Obligatory vector

HD Header vector

The Header vector HD is a fixed-length vector of 12 bytes, which stores the Beilstein Registry Number, the size of the CT in bytes, number of atoms, number of optional lists, etc.

3.2 Obligatory lists

PI Pi-bonding electron list

The Pi-bonding electron list PI specifies for each atom in the structure graph the number of valence electrons which contribute to one or more ¶-bonds.
The PI list always follows the HD vector in the CT.

FR From list

The From list FR specifies for each atom in the structure graph the index of the lowest-indexed atom to which the current atom is attached. For atoms which are attached to two or more lower-indexed atoms, the attachments to atoms other than the lowest-indexed atom constitute ring-closures, and are specified in the RC list.

The FR list always follows the PI list in the CT.

3.3 Optional lists

Each optional list has a code number. the optional lists follow the FR list in ascending order of this code.

RC Ring-closure list

The Ring-closure list RC specifies for each atom in the structure graph which is attached to two or more lower-indexed atoms the index of those attached atoms other than the lowest-indexed atom.

AT Atom list

The Atom list AT specifies the location of non-Carbon atoms in the structure graph. Special conventions govern when Hydrogen atoms are included in the structure graph.

LH Localised Hydrogen list

The Localised Hydrogen list LH specifies the number of non-tautomeric Hydrogen atoms, not including its isotopes D and T, attached to each atom in the structure graph.

AS Stereo-atom list

The Stereo-atom list AS specifies the absolute configuration of each chirality centre (centre of asymmetry) and of each centre of pseudo-asymmetry in the structure graph.

The codes are assigned by application of the Cahn-Ingold-Prelog Sequence Rule System to the ligands of the asymmetric centre. A pseudo-asymmetric centre is defined as an asymmetric atom of which two ligands differ only in their chirality.

BS Stereo-bond list

The Stereo-bond list BS specifies the absolute configuration of each stereogenic double bond in the structure graph.

XS Stereo-axis list

The Stereo-axis list XS specifies the sense of chirality of each chirality axis in the structure graph.

VA Non-default valence list

The Non-default valence list VA specifies the atoms in the structure graph which have a non-default valence.

LC Localised charge list

The Localised charge list LC specifies the position of localised charges in the structure graph, excluding those which are tautomeric or delocalised.

DC Delocalised charge list

The Delocalised charge list DC specifies the number and type of delocalised charges, and the atoms of the structure graph over which each is delocalised.

LR Localised unpaired valence electron (radical) list

The Localised unpaired valence electron (radical) list LR specifies the atoms in the structure graph which carry a localised unpaired valence electron.

DR Delocalised unpaired valence electron (radical) list

The Delocalised unpaired valence electron (radical) list DR specifies the atoms in the structure graph over which one or more unpaired valence electrons are delocalised.

MA Abnormal mass (known location) list

The Abnormal mass (known location) list MA specifies non-Hydrogen atoms in the structure graph which have an abnormal mass.

MU Abnormal mass (unknown location) list

The Abnormal mass (unknown location) list MU specifies the atomic number of each non-Hydrogen element in the structure graph with which an abnormal mass at unknown location is associated.

HI Hydrogen isotope (known location) list

The Hydrogen isotope (known location) list HI specifies the atoms in the structure graph to which are attached one or more isotopes of Hydrogen which

(i) are not specified in the structure graph, and

(ii) are not tautomeric.

HU Hydrogen isotope (unknown location) list

The Hydrogen isotope (unknown location) list HU specifies the atomic mass of each isotope of Hydrogen present at unknown location.

TM Tautomer group (mobile) list

The Tautomer group (mobile) list TM specifies the tautomeric groups and the atoms in the structure graph to which each may be attached. The tautomer groups are recognised algorithmically.

TL Tautomer group (localised) list

The Tautomer group (localised) list TL specifies the tautomeric groups and the particular atom in the structure graph to which each is attached in the Registry display formula. This list is used to display a tautomeric Registry compound in its preferred tautomeric form.

SD Supplementary descriptors list

The Supplementary descriptors list SD contains supplementary descriptor codes and text strings which describe structure attributes which cannot be described elsewhere in the CT.

4. The Intermediate File.

After input, the data are loaded from the floppy disks into an intermediate file on the mainframe. The data are run through several plausibility checks and when necessary are corrected. In this intermediate file all data from the various input sources are converted to the same file- and data-structure according to our data structure definition.

5. The Database.

After having been converted to the same file structure the data will be loaded into an Adabas-managed database (Fig. 2). More than 10 different commercial database

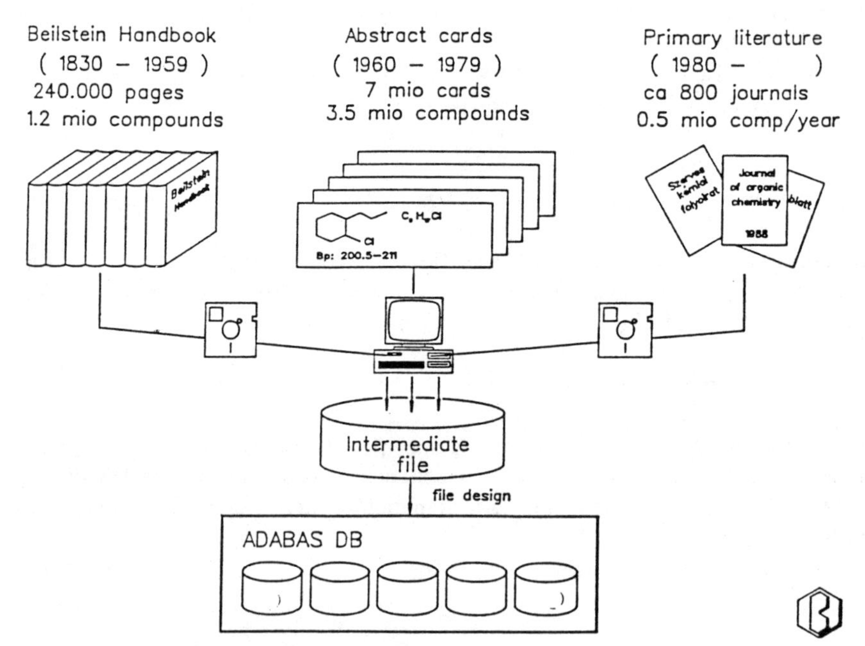

Figure 2. The data, once converted to the same file structure, will then be loaded into an Adabas-managed database.

management systems have been evaluated during our systems analysis phase in 1984/85. The two highest scoring systems have been benchmarked with an artificial BEILSTEIN-structured database with 200,000 compounds. Adabas scored best in practically all tests (loading, updating, retrieval etc.).

An Adabas-based update- and retrieval-system is currently being developed in our Institute together with a West-German software house. This system will be used to make final corrections to the database, to append compounds and retrieve compounds for checking, and for writing Handbook articles. It will also be licensed to inhouse customers (see below) of the BEILSTEIN-database. There will be a micro- and a mainframe-based version of this software.

6. Database Search Possibilities.

The data structure of the BEILSTEIN database allows many different accessing methods to the information dependent only on the retrieval software used. The following access methods are available with our inhouse retrieval software:

- Via a graphical or alphanumerical input - depending on the type of computer terminal (structure or substructure search).

- By searching for numerical terms (physical data, keywords or boolean terms, such as the existence of spectra, etc.) [7]. These terms can be searched separately or in boolean combinations ("and", "or", "not").

- By combination of the two search methods described above.

- By searching other keyfields, such as the molecular formula, the CAS Registry Number or the molecular, structure-related Beilstein Registry Number (Beilstein Prime Key).

7. Access to the Beilstein Database.

The Beilstein Database can be searched online on several Online Hosts:

- Since December 1988, Beilstein has been available online on STN. The database is physically located at the STN-Node in Karlsruhe but can be searched from any place in the world with access to STN. Special numeric search capabilities have been introduced to the Messenger Search System to allow sophisticated searching of all Beilstein data fields (see also the other papers of this book). Structure and substructrue searching is done with the same commands as the CAS Registry File. Stereospecific searches are not possible at this time on STN, because this feature is not yet available with the Messenger search software.

- Beilstein went online on Dialog in the last quarter of 1989. Dialog also implemented very powerful new techniques for numeric searchung. In addition, Dialog implemented S^4, the substructure search system jointly developed by Beilstein and Softron Corp. [6]. From the second quarter of 1990, this system will allow stereospecific structure and substructure searches on Dialog. S^4 is also available for inhouse searching on private hosts or companies.

- Beilstein will also go online on Orbit/BRS (Maxwell Online) in the middle of next year. No details of the implementation are available at this time.

- It is also planned to implement the Beilstein Database on the Swiss Online Host Datastar in the near future.

Inhouse Access to the Structure File of the Database is also available. The complete structure file including Beilstein Registry Numbers and references can be licensed for an annual license fee. Licensing is possible with or without our substructure search software package S^4. The Inhouse license allows unlimited structure and substructure searching of the file. The factual data of the target compounds can be retrieved through online hosts via the Beilstein Registry Number.

8. References

1. Prof. Dr. Reiner Luckenbach, Dr. Clemens Jochum, Beilstein-Institut für Literatur der Organischen Chemie, Frankfurt am Main, Achema-Jahrbuch 88, Band 1: Forschung und Lehre des Chemie-Ingenieurswesens, Seite 245-247

2. Clemens Jochum, "Building a structure-oriented numerical Factual Database", in: Chemical Structures, W. A. Warr (Hrsg.), Springer Verlag, 1988, Seite 187-193

3. S. M. Welford, "Tautomer Processing in the Beilstein Registry System", in: Software-Entwicklung in der Chemie 2, J. Gasteiger (Hrsg.), Proceedings des Workshops Computer in der Chemie, Hochfilzen /Tirol, Springer-Verlag, 1988, Seite 35-43

4. L. Domokos, "Data About Data", in: Physical Property Prediction in Organic Chemistry, C. Jochum, M.G. Hicks, J. Sunkel (Hrsg.), Proceedings of the Beilstein Workshop, Schloß Korb, Italy, Springer-Verlag, 1988, Seite 11-19

5. S. M. Welford, "Die Datenstruktur des Beilstein für organische Verbindungen", in: Software-Entwicklung in der Chemie 1, J. Gasteiger (Hrsg.), Proceedings des Workshops Computer in der Chemie, Hochfilzen/Tirol, Springer-Verlag, 1987, Seite 5-11

6. M.G. Hicks, "Substruktursuche: Ein Leistungsvergleich von DARC, HTSS, MACCS und S4 durch das Beilstein-Institut", in: Software-Entwicklung in der Chemie 3, G. Gauglitz (Hrsg.), Proceedings des Workshops Computer in der Chemie, Hochfilzen/Tirol, Springer-Verlag, 1989, Seite 9-23

7. J. Sunkel, "Neue Suchstrategien in Numerischen Organischen Faktendatenbanken", in: Software-Entwicklung in der Chemie 3, G. Gauglitz (Hrsg.), Proceedings des Workshops Computer in der Chemie, Hochfilzen/Tirol, Springer-Verlag, 1989, Seite 25-29

RECEIVED May 18, 1990

Chapter 3

STN Implementation of Factual and Structure Databases

Andreas Barth

STN International, FIZ Karlsruhe, D-7514 Eggenstein–Leopoldshafen 2, Federal Republic of Germany

Since December 1988, the first part of the Beilstein database has been available online through STN International. This file covers a broad scope of chemical and physical information from 1830 to date. In this paper an overview of the design and implementation of this file on STN will be presented. The design of the database is briefly discussed. Following, a description of Chemical Substance Identification, Chemical Reaction Information, Physical Property Data, General Fields and Bibliographic Data is given. It is illustrated that the implementation of the database on STN allows for rather sophisticated searches of various properties, including chemical reaction information.

The first part of the Beilstein database of organic substances was introduced in December 1988 on STN International. It is the first time that a factual database containing such a large number of physical and chemical entities has become publicly available as an online database. Initially, the database contained the structures and factual data of approximately 350,000 heterocyclic substances from the printed handbook, covering the time span from 1830 to 1959. Since then the number of substances has increased substantially and is expected to reach approximately 3.5 mio. organic

0097–6156/90/0436–0024$06.00/0

compounds by March 1990. The database will then contain updates through 1980.

In the meantime the Beilstein Institute has begun the extraction of the primary literature starting from 1980. This data is directly input into personal computers using menu-driven excerption programs developed by the Beilstein Institute and Softron GmbH. After a short error checking and reviewing process, the data will be loaded immediately into the online database. Within a few years the Beilstein database will include all the excerpts from the current literature and will be almost up-to-date (i.e the data will be loaded into the database within less than one year).

There are four different sources contributing to the online database (see also Chapter 2):

- Beilstein handbook (currently up to 1959),
- Literature excerpts on file cards (1960 - 1979),
- Beilstein handbook (1959 - 1979, in print),
- Literature excerpts in machine-readable form (from 1980).

Only the data from the handbook is critically reviewed. In the database this is indicated by the note 'Handbook Data'. The first source for the online database is the famous Beilstein Handbook of Organic Chemistry, the largest collection of critically reviewed data of organic chemistry. It was published for the first time in 1918 and is now available in the 4th edition, consisting of the Basic Series and four Supplementary Series. This covers the complete published literature on organic chemistry through 1959. Recently, the first volumes of the fifth Supplementary Series have been delivered, and the complete set of volumes will be continuously printed during the next decade(s). Each organic substance is identified by a chemical name given in IUPAC-oriented nomenclature and a structure diagram. In addition, a large set of physical properties and chemical information is described together with the corresponding literature references. The scope of information may cover (1): substance identification information, synthesis and reaction data, structure and energy parameters, state of aggregation, mechanical properties, thermodynamic data, transport phenomena, optical and spectral data, magnetic and electrical data, electrochemical behaviour, and multi-component system data. In analogy to the printed Handbook, the documents in the Beilstein database are substance-oriented (i.e. all factual information is associated with a well-defined chemical substance and structure).

For the individual chemical substances, the number of associated factual data may vary significantly. The minimum information corresponding to a substance comprises the identification data and one physical or chemical entity. For pyridine, currently the most comprehensive substance, all factual data are available and an offline print consists of more than 540 pages. With each factual data field, there is at least one literature reference and sometimes an additional note giving further information. Statistics on the number of substances per entity is given in Figure 1 as a bar chart.

These statistics are taken from the current database of 1,745,686 substances with 460,846 records from the handbook file (heterocyclic and acyclic substances) and 1,284,840 records from the excerpt file (heterocyclic substances). It can be seen from this figure that the major part of factual data is comprised of preparation data (PRE, 87.4 %), reaction data (REA, 10.2 %), melting point (MP, 68.2 %), and boiling point (BP, 12 %). While this may be a result of the content of the current file, it is not expected that these figures will vary significantly when the file contains a more balanced set of substance information from the handbook and the excerpts.

Design of the Database

According to the data structure of the Beilstein database, there are several hundred search fields and more than 140 different display formats (2, 3). As shown in the first section, the type and amount of information which is available for a particular substance may vary significantly. To obtain an overview of the available fields for a given substance the display format FA (**F**ield **A**vailability) can be used. In Figure 2, the table of content for the substance tryptophane is shown. The Messenger command language used here is described elsewhere (4).

In the first column of this table the display formats are given, the full name is shown in the second column and the number of occurrences is displayed in the last column. In this case, we find that there are 2 occurrences of preparation. The number of occurrences is a direct indication of the number of different preparation methods for the substance. To display the data for the substance, one may simply use the codes from the first column of Figure 2. A display of IDE (**Ide**ntification of Substance including BRN + CN + MF + SO + FW + LN + STR) and PRE (Preparation) is shown in

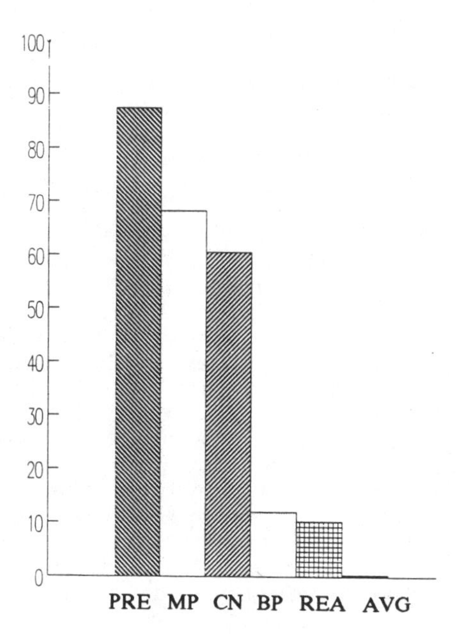

Figure 1. Percentage of database records containing certain properties.

=> SEARCH tryptophan/CN

L9 1 TRYPTOPHAN/CN

=> DISPLAY FA

L9 ANSWER 1 OF 1

Code	Field Name	Occur.
MF	Molecular Formula	1
CN	Chemical Name	1
FW	Formula Weight	1
SO	Beilstein Citation	1
LN	Lawson Number	1
NTE	Notes	1
PRE	Preparation	2
MP	Melting Point	6
REA	Chemical Reaction	15
RSTR	Related Structure	3
INP	Isolation from Natural Product	6
CPD	Crystal Property Description	2
ORP	Optical Rotatory Power	4

Figure 2. Table of Contents for Tryptophane.

Figure 3. It should be noted that any combination of formats including combined (predefined) and custom formats is allowed.

For a deeper understanding of the Beilstein database, it is necessary to outline the structure of a Beilstein document. There are essentially four different information levels (see Figure 4). All the substance identification information comprises the first level. It is actually associated with a registered Beilstein compound (title compound). On the second level, one finds the availability information. This comprises the search fields Field Availability (FA), Property Hierarchy (PH), Controlled Terms (CT), and Controlled Terms of Multi-Component Systems (CTM). The content of these fields indicates whether there is information available for a specific property. The factual information, i.e. numeric values of properties and reaction information, are found on the third level (measurement). On the fourth level the bibliographic information is given referring to the measurements of the next higher level.

There is a one-to-many relationship between the higher and the next lower level in this hierarchy. This means that for a given substance, there could be many properties available, for each property there could be several measurements, and for a measurement there could be more than one citation. A search could be performed in fields of different levels. In any case, the answer set will consist of Beilstein Registry Numbers (BRN) plus some additional information. This means that the result of a search is always a Beilstein title compound. The information about the search level can be reconstructed from the additional information in the answer set and is used for the various display formats. A certain value of the (P)-proximity is assigned to each measurement (instance of data) and the individual references are associated with a particular (S)-proximity value.

All information of the first two levels is indexed in standard search fields, e.g. CN (Chemical Name) or MF (Molecular Formula). The factual data has a slightly different structure since the main entity is generally depending on further parameters. As an example, the data structure of Dipole Moment is shown in Figure 5. Here, the main entity is depending upon the parameters Temperature, Method, and Solvent. The values for the Dipole Moment and the corresponding Temperature are given in Debye and degree Celsius, respectively. The (P)-operator allows to perform a search of the Dipole

```
=> display ide pre

L1   ANSWER 1 OF 1

BRN  86196  Beilstein
MF   C11 H12 N2 O2
CN   Tryptophan
FW   204.23
SO   0-22-00-00546; 1-22-00-00677; 0-22-00-00550; 4-22-00-06765; 5-22
NTE  stereoisomeres of unknown configuration.
LN   27812

          H
          N

                  CH2CHCO2H
                     |
                    NH2

Preparation:
PRE
     Educt:  casein
     Detail: Reaktion ueber mehrere Stufen
     Reference(s):
     1. Organic Syntheses, 10 <New York 1930>, S. 100
     2. Dakin, Biochem.J. 12 <1918>, 302, CODEN: BIJOAK
     3. F. Hoppe-Seyler, H. Thierfelder, Physiologisch- und
        pathologisch-chemische Analyse, 9. Aufl. <Berlin 1924>, S. 313
     4. Abderhalden, Chem.Ber. 42 <1909>, 2333, CODEN: CHBEAM
     5. Abderhalden, Kempe, Hoppe-Seyler's Z.Physiol.Chem. 52 <1907>,
        208, CODEN: HSZPAZ
     6. Hopkins, Cole, J. Physiology 27 <1901>, 420
        J. Physiology, 29 <1903>, 453
     Note(s):
     7. Handbook Data
     8. l-tryptophan
```

Figure 3. Display of data for Tryptophane.

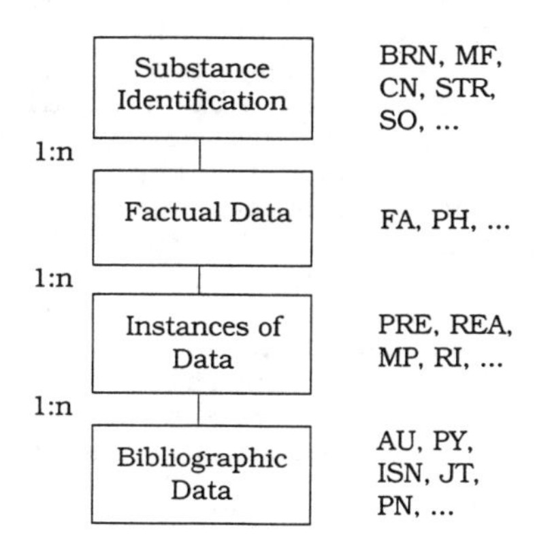

Figure 4. Document structure and information levels.

Moment at specific parameter values. The main qualifier (DM) is also used to display the data for Dipole Moment.

In addition to the custom formats consisting only of a single entity, there is a hierarchy of combined (predefined) display formats. These enable the user to display almost any amount of data using a single display format. There is also a dynamic display feature which shows the customer the data related to his/her search query plus the substance identification information (QRD = **Q**uery-**r**elated **D**ata). To display the content of a Beilstein document, the Field Availability (FA) is used (see Figure 2).

In the Beilstein database, the information is given in different forms, as textual data (free text), keywords, numeric values (ranges), and structures. Free text and keywords can be searched in the same way as in bibliographic databases. Numeric values are searched as ranges (see chapter 8) using the numeric relation operators, e.g. '<' (less than) or '>' (greater than). Chemical structures can be searched as exact structures or as substructures. They can be build either offline using a graphical structure editor or online using the Messenger STRUCTURE command. Examples for online searches are presented in the next sections.

Chemical Substance Identification

All substances in the Beilstein database are identified by a sequential **B**eilstein **R**egistry **N**umber (BRN). In addition, there is a Chemical Name (CN) and Synonyms (SY), a Molecular Formula (MF) and related formulas, a Formula Weight (FW), and a Structure (STR). All these fields are both searchable and displayable. Furthermore, there are many additional search fields generated from these input fields. Chemical Names are given in IUPAC-oriented nomenclature. They are indexed as complete names in CN and as parsed segments in the fields CNS (Chemical Name Segments) and BI (Basic Index). The segments are generated using two different algorithms: 1. by parsing the names at all special characters like hyphen and comma, and 2. by applying a dictionary of natural segments developed by the Beilstein Institute. Chemical name segments can be searched using the well-known proximity operators (S), (W), and (A). An example for such a search is given in Figure 6. Here, we are searching for derivatives of salicyl aldehyde. The answer set comprises 59 hits and the 4th answer is also shown in Figure 6. A hit resulting from a search in the Basic Index may also stem from a starting material of a reaction or from a by-product. These names are also

Field Name	Field Qualifier	Unit
Dipole Moment	DM[1]	D
Temperature	DM.T	CEL(°C)
Method	DM.MET	-
Solvent	DM.SOL	-

Figure 5. Design of Physical Entities: Dipole Moment

```
=> search salicyl (w) aldehyde/cns
           147 SALICYL
          2814 ALDEHYDE/CNS
L1          59 SALICYL (W) ALDEHYDE/CNS

=> display 4


L1  ANSWER 4 OF 59

BRN  351066  Beilstein
MF   C25 H26 N2 O
CN   salicylaldehyde-<2-(3,4-dihydro-1H-<2>isoquinolylmethyl)-phenethyl
     imine>
     Salicylaldehyd-<2-(3,4-dihydro-1H-<2>isochinolylmethyl)-phenaethyl
     imin>
FW   370.49
SO   2-20-00-00179
LN   24291; 14535; 8629
```

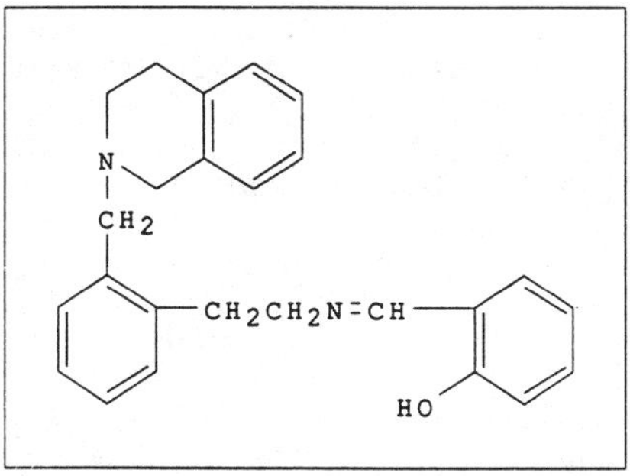

Figure 6. Example Search Using Chemical Name Segments

indexed in the Basic Index, but they are not necessarily registered Beilstein compounds.

The Molecular Formula (MF) and the associated search fields are another possibility to identify a chemical substance. A number of additional search terms is generated from the molecular formula (see Figure 7). At first, the Molecular Formula is indexed in MF and BI (Basic Index). In addition, the Single Atom Counts are generated for all chemical elements plus some pseudo atom counts like X (Halogen Atoms) and M (Metal atoms). For each element the corresponding periodic group and element group is created. Furthermore, a total Element Count (ELC), a total Atom Count (ATC), and the Element Symbols (ELS) are indexed. The latter fields are especially useful to limit the search to certain ranges of atoms or elements.

Chemical structures are the most important key to identify substances in the Beilstein database. The user interaction and the substructure search capabilities are identical to those of the CAS Registry database and are described in detail elsewhere (5). In Figure 8, a substructure search for derivatives of adenine is shown. Using PC-based software, like STN Express or Molkick, it is also possible to build structures offline and upload the connection tables into the database.

It is also possible to use the Lawson Number (LN) for searches of substance in Beilstein. This number is a fragment code based on the Beilstein system. Using this number the customer may perform similarity searches or browse through a set of substances. The possibilities to use the Lawson Number are described in detail in Chapter 10 of this book.

Chemical Reaction Information

Although the Beilstein database is not a typical reaction database, there is a large amount of chemical reaction information available. It is possible to find data on substance preparation, chemical behaviour and isolation from natural products (biosynthesis). Currently, all searches must be performed as text searches for chemical name segments or as searches for the availability of data. This is certainly a limitation for reaction searches. In many cases, one retrieves the complete reaction information, including the literature references. There are some cases, however, which consist only of a reference.

Using the Field Availability (FA), one could search for the availability of reaction data. In the following example we are interested in the biosynthesis of flavone

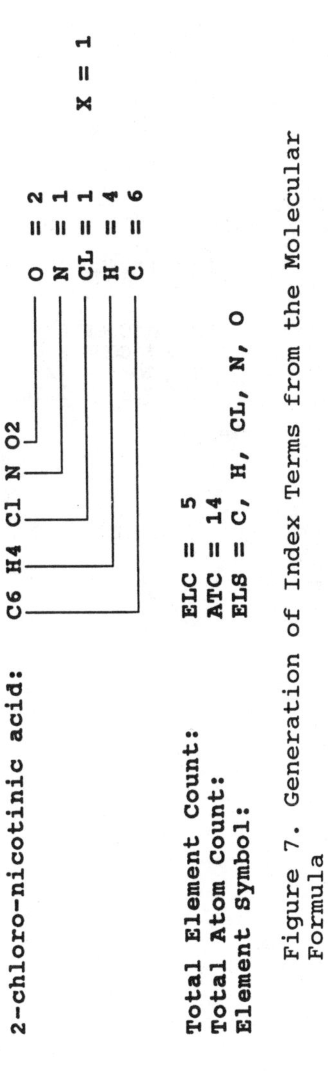

Figure 7. Generation of Index Terms from the Molecular Formula

```
=> search adenine/cn
L2           1 ADENINE/CN

=> display brn


L2  ANSWER 1 OF 1

BRN  5777  Beilstein

=> str 5777
:end
L8   STRUCTURE CREATED

=> s 18 sss sam
SAMPLE SEARCH INITIATED 17:15:35
SAMPLE SCREEN SEARCH COMPLETED -      94 TO ITERATE
 54.3% PROCESSED      51 ITERATIONS                          50 ANSWERS
INCOMPLETE SEARCH (SYSTEM LIMIT EXCEEDED)
SEARCH TIME: 00.00.05

FULL FILE PROJECTIONS:  ONLINE  **COMPLETE**
                        BATCH   **COMPLETE**
PROJECTED ITERATIONS:       650 TO     1230
PROJECTED ANSWERS:          634 TO     1208

L9           50 SEA SSS SAM L8

=> d


L9  ANSWER 1 OF 50

BRN  1231163  Beilstein
MF   C11 H14 N8 O4
FW   322.28
SO   5-26
LN   30692; 20559
```

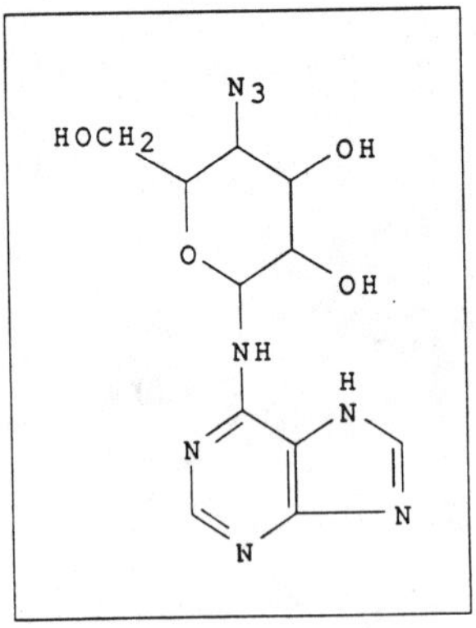

Figure 8. Substructure search for derivatives of Adenine.

derivatives. Information on biosynthesis could be searched in the field Isolation of Natural Products (INP). Since we are not looking for any specific synthesis, we can search for 'INP' in FA and combine this with the name segment 'flavon?' in CNS (see Figure 9). The question mark ('?') stands for a right truncation of the search term.

The names of the substances taking part in the preparation of a compound are indexed in individual subfields of Preparation (PRE). The reaction scheme for substance preparation is given in Figure 10. Here, a compound C is built from the starting materials A and B. The title compound C could be identified by BRN, CN, STR etc, the starting materials are indexed in PRE.EDT or PRE.RGT and by-products are found in PRE.BPRO. An example for a search of preparation information is shown in Figure 11.

In the Beilstein database, one may also search for the Chemical Behaviour (REA) of a compound. The corresponding reaction scheme is shown in Figure 12. The title compound A is again identified by a Beilstein Registry Number (BRN), a Chemical Name (CN), and a Structure (STR) etc. Substance B is called the Reaction Partner (REA.RP) or Reagent (REA.RGT) and the Products C and D are found in REA.PRO. The strategy for searching this type of information is completely analogous to the case of substance preparation (see Figure 11).

All searches discussed so far retrieve the compounds from the complete Beilstein database, including both the handbook data and the excerpts (unchecked data). Thus the user gets a mixture of evaluated and non-evaluated data. However, it is also possible to restrict the searches to either group of information. There is a keyword subfield associated with each factual data group (entity), containing the keyword 'Handbook' or 'Unchecked'. An example for a search of chemical behaviour restricted to evaluated (handbook) data is presented in Figure 13. The display format HIT shows only reactions which have been evaluated by the Beilstein Institute.

It is important to note that it is not always obvious to an online searcher where the reaction information is indexed in the Beilstein database. Sometimes it is necessary to execute a query both in the preparation and the reaction fields. It is also necessary to keep in mind that only a subset of the Beilstein compounds have reaction data available.

```
=> SEARCH flavon?/CNS AND inp/Fa

=> s flavon?/cns and inp/fa
            1378   FLAVON?/CNS
           19693   INP/FA
L2            45   Flavon?/CNS AND INP/FA
```

Figure 9. Search for the Biosynthesis of Flavone Compounds

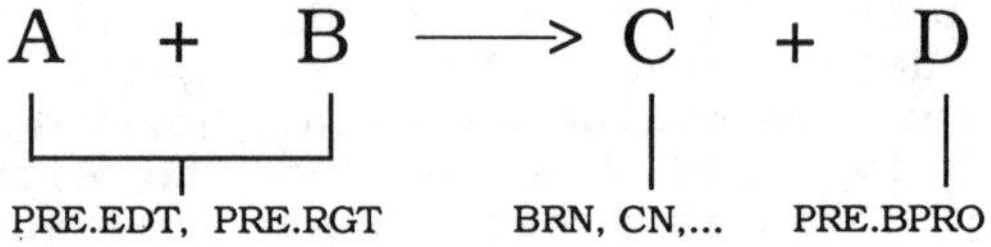

A + B ——> C + D

PRE.EDT, PRE.RGT (A, B) BRN, CN,... (C) PRE.BPRO (D)

A, B: Educts, Reagents, Solvents, Catalyst

C: Title Compound

D: By-Product

Figure 10. Reaction Scheme for Substance Preparation

```
=> SEARCH   uric(W)acid/CNS AND
            barbituric(W)acid/PRE.EDTD

       464  URIC
     91543  ACID/CNS
       124  URIC(W)ACID/CNS
      3414  BARBITURIC
    115569  ACID/PRE.EDT
      1396  BARBITURIC(W)ACID/PRE.EDT

L2       6  URIC(W)ACID/CNS AND
            BARBITURIC(W)ACID/PRE.EDT
```

Figure 11. Search Strategy for Substance Preparation Data

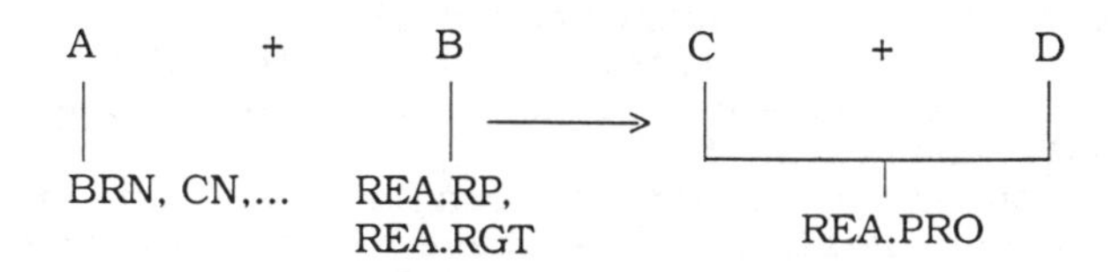

A: Title Compound
B: Reaction partner(s), reagents, solvents, catalyst
C, D: Products

Figure 12. Reaction Scheme for Chemical Behaviour

```
=> search thiophene/cns and handbook/pre.kw
          4728 THIOPHENE/CNS
        426150 HANDBOOK/PRE.KW
L4        4462 THIOPHENE/CNS AND HANDBOOK/PRE.KW

=> display hit


L4 ANSWER 1 OF 4462

Preparation:
PRE
     Educt:  tetraphenyl-<1,4>dithiin
     Reag:   peroxyacetic acid
     Reference(s):
     1. Kirmse, Horner, Liebigs Ann.Chem. 614 <1958> 4, 18, CODEN: LACHDL
     Note(s):
     2. Handbook Data
PRE
     Educt:  tetraphenylthiophene
     Reag:   hydrogen peroxide, acetic acid
     Reference(s):
     1. Hinsberg, Chem.Ber. 48 <1915>, 1612, CODEN: CHBEAM
     Note(s):
     2. Handbook Data
```

Figure 13. A Limited Search for Evaluated (Handbook) Data

Physical Property Data

The physical properties and their retrieval is discussed in detail in Chapter 8 of this book. Here, we discuss only shortly the main ideas of physical property retrieval.

One can search for the presence of physical properties in one of the following fields: Field Availability (FA), Controlled Terms (CT), Controlled Terms of Multi-Component Systems (CTM) or Property Hierarchy (PH). A property name is indexed in FA only if numeric values are present for this entity. In CT and CTM, there are only entries if there is only a reference but no numerical value. In both cases, one finds an entry in PH, i.e. it is a 'sum' of both FA and CT/CTM. Figure 14 shows an expand list of the Property Hierarchy (PH). Here, we are looking for critical properties. The Expand list shows 8 different properties (E4 - E11). A search for the presence of data is like a search in a very simple thesaurus file.

Most physical properties can be searched as numeric ranges in the Beilstein database. The property values are indexed in specific units and are generally associated with a set of parameters. As an example for the design of a physical property, the structure of the Dipole Moment (DM) is shown in Figure 5. To search for a property at a given value of a parameter, one must combine the fields with the (P)-proximity. In Figure 15, we search for substances having a Heat of Formation (HFOR) greater than 1.000 J/mol at a Temperature (HFOR.T) of 20 °C. The answer set comprises all substances with a Heat of Formation overlapping the search range.

Except for a few dimensionless constants, all physical properties are measured in a particular unit. All values of a given property are indexed in the same unit. To see the set of physical units for the database one has to specify HELP UNITS at a Messenger arrow ('=>') prompt. The EXPAND list for a field also shows the physical unit. With the December 1989 release of the Messenger software, the user can work in his own set of units and the software automatically converts the query into the units of the property on the file.

Analogous to the chemical reaction retrieval, there is also the possibility to distinguish between evaluated (handbook) and non-evaluated (unchecked) data using the keyword subfields. These searches are described in more detail in Chapter 8 of this book.

General Fields and Bibliographic data

In addition to the entities described above, there are several fields with general content. First, there is the Basic Index (BI) containing data as single words from all text fields. In BI one can search for chemical name segments or for general information as toxicity data or

```
=> expand critical/ph

E1        105544      CPD/PH
E2           110      CRD/PH
E3             0 -->  CRITICAL/PH
E4            34      CRITICAL DATA FOR MIXTURES/PH
E5           110      CRITICAL DENSITY/PH
E6            30      CRITICAL FREQUENCY (OR WAVELENGTH)/PH
E7            40      CRITICAL MICELLE CONCENTRATION/PH
E8           165      CRITICAL PRESSURE/PH
E9           174      CRITICAL SOLUTION TEMPERATURE/PH
E10          252      CRITICAL TEMPERATURE/PH
E11           37      CRITICAL VOLUME/PH
E12          165      CRP/PH
```

Figure 14. Expand List of the Property Hierarchy

```
=> s hfor > 10000 (p) hfor.t = 20
             89 HFOR > 10000
             14 HFOR.T = 20
L20           5 HFOR > 10000 (P) HFOR.T = 20

=> d hit


L20 ANSWER 1 OF 5

Enthalpy of Formation:
Value (HFOR)    Temp.(HFOR.T)    Press.(HFOR.P)   Ref.  Note
(J/mol)         (Cel)            (Torr)
----------------+----------------+----------------+-----+-----
19259           20.0                              2     1

Reference(s):
2. Sunner, Acta Chem.Scand. 11<1957>1766, 1770, CODEN: ACSAA4

Note(s):
1. Handbook Data
```

Figure 15. Example for a Numeric Property Search

ecological information. This field can be used in a similar way as in other files.

The factual data are always given together with one or more literature reference(s). A part of this bibliographic data is also searchable in the file. Since Beilstein quotes only the surname of the first author there is a limited way to search names in AU. The **P**ublication **Y**ear is searchable in PY, the **P**atent **N**umber is found in PN, the **J**ournal **T**itle is given in JT and the corresponding CODEN is indexed in ISN (**I**nternational **S**tandard **N**umber). It is important to note that the bibliographic information is referring to an instance of a measurement of a property or a chemical reaction (see Figure 4). This information corresponds to the 'lowest' information level and not to a title compound as a whole like the reference number in the Source (SO) field. The (P)-proximity is used to refer to an instance of factual data, however, if two or more bibliographic search terms are combined in one query, the (S)-proximity must be used. The reason for this is simply that there may be more than one reference for a single instance of factual data.

The substances in the Beilstein database are identified by the Beilstein Registry Number (BRN). This is a sequential number assigned to each well-defined organic substance. Since the registration process of Chemical Abstracts Service and Beilstein Institute are different it is difficult to identify substances in the two databases. To overcome this problem, the two database suppliers have finally agreed to register the Beilstein substances at Chemical Abstracts Service. The CAS Registry Numbers (RN) are assigned to all Beilstein substances which have already an RN; new substances are not registered. Since the registration concepts are different, the correspondence will be a one-to-many, i.e. a given Beilstein substance may have one or more CAS Registry Numbers. Chemical Abstracts Service will assign a locator flag for all registered Beilstein substances.

Conclusion

The Beilstein database is the largest numerical database both with respect to substances and properties. It is also an important source for reaction information especially for the early area of chemistry. At the end of the first quarter of 1990, the Beilstein database will comprise about 3.5 million organic substances, covering the time period from the beginning of Chemistry until 1980. With the multi-component systems, there will be another 0.5 million compounds to be added to the file in 1990. Following this, it can be assumed that there will be large a overlap between the Beilstein and the CAS Registry file. However, the concepts of Chemical Abstracts Service and Beilstein Institute are different. While the latter is focussing on factual (numeric) data, Chemical Abstracts Service is concentrating on chemical concepts and bibliographic information. Therefore, the two databases should be viewed as complementing each other. Since the registration of the Beilstein substances at Chemical Abstracts Service enables the customers to perform crossfile searches in both directions, it is now possible to work with both databases in a very easy and user-friendly way.

Acknowledgments

The funding of this work by the Federal German Ministry for Research and Technology is greatly acknowledged.

Literature Cited

1. What is Beilstein?, Springer-Verlag

2. Beilstein Database Description Manual, STN Karlsruhe, December 1988

3. Online Searching on STN: BEILSTEIN Reference Manual, S.R. Heller and G.W.A. Milne, Springer-Verlag, 1989

4. A Guide to Commands and Databases, STN Columbus, October 1988

5. Using CAS ONLINE, The Registry File, STN Columbus, June 1985

RECEIVED May 17, 1990

Chapter 4

An Overview of DIALOG

Ieva O. Hartwell[1] and Katharine A. Haglund[2]

[1]Consultant, 2908 Georgetown Drive, Midland, MI 48640
[2]Dialog Information Services, Palo Alto, CA 94304

This chapter gives an overview of the implementation of the *Beilstein* database on Dialog Information Services, Inc.'s DIALOG system with an emphasis on the indexing, search and display features that are unique to DIALOG. DIALOG has made virtually every word and piece of factual data searchable giving the user unprecedented access to the wealth of information in the *Beilstein Handbook*. Chemical substance names have been segmented to provide greater access to substances by functional groups. Related terms have been incorporated to assist the user in searching those data fields that are currently provided in German as well as in searching non-numeric properties. Special collected data fields have been created to minimize the number of search codes a searcher has to use to search physical properties with dependent parameters. Additional index entries have been created from molecular formulas to permit generic searching based on the elemental composition of a substance. Preparations and chemical reactions can be searched very precisely by specifying components and conditions or very broadly by specifying that the terms occur anywhere in a reported preparation or reaction. Structure searching is based on the S4 system produced by SOFTRON GmbH and uses the new set of DIALOG QS (Query Structure) commands with ROSDAL (Representation of Organic Structure Description Arranged Linearly) graphic or text structure input.

BEILSTEIN ONLINE (File 390 on DIALOG) is the online version of the *Beilstein Handbook of Organic Chemistry* and more recent data that is yet to be included in the published *Handbook*. The *Handbook* data consists of facts on organic compounds that have been collected and critically reviewed by Beilstein Institute staff for internal consistency, accuracy and significance from literature published around the world since 1779. *Handbook* data corresponds to the main work and four supplementary series covering the literature through 1959. The *Handbook* is organized by structure of the compound into 27 "volumes," each of which may be one or more physical books.

0097–6156/90/0436–0042$06.50/0

The more recent data, currently being collected, reviewed and evaluated for inclusion in the fifth supplement, is known within the online file as "Short File" data and covers the literature published from 1960 through 1979. Both new data on existing compounds and new data on new compounds are included as Short File data. Except for five basic physical properties (melting point, boiling point, refractive index, density, and optical rotatory power), Short File data is presented as keywords with literature references.

Each organic compound in the database has a single online record that includes all *Handbook* data from the main work and various supplements plus Short File data, if appropriate. Each record also includes the *Beilstein* citations for the location of the factual data in the printed *Handbook*.

TEXT SEARCHING

Virtually every term and piece of factual data in BEILSTEIN ONLINE is searchable in either the Basic Index or one of the Additional Indexes or both.

Basic Index. The Basic Index includes all terms of subject importance. These terms are phrases, words and chemically significant word segments extracted by the DIALOG segmentation process. The segmentation process is described in a later section. Because of the number of different major subject areas covered by the fields in the Basic Index, the user has the option to limit a search to a specific subject using field "suffixes." These suffixes are summarized in Table I. Suffixes may correspond to individual data fields or to collected fields of like data. For example, the name of the substance of record is indexed in the individual fields /CN, /GC or /SY, as available, and the collected field /NA, which encompasses these three individual fields. In addition, substance names may occur in most of the other Basic Index fields as reagents, partners, solvents, products, by-products, etc. Therefore, to find a substance of record searching by name, the /NA suffix can be used to restrict the search to only the Chemical Name and Synonyms field for the substance of record excluding other occurrences of the name in the Basic Index.

Chemical Names. Chemical names in all Basic Index fields, except Comment and Isolation from Natural Product fields, have been indexed by phrase, word and segment. Comment and Isolation from Natural Product fields are indexed by word and segment.

Systematic chemical names are provided for *Handbook* substances. They are based on IUPAC (International Union of Pure and Applied Chemistry) nomenclature rules and are present in the Chemical Name (in English) and German Chemical Name (in German) fields of the Basic Index. Synonyms are primarily present for Short File substances and may be in German or in English. Chemical names present in the other fields of the Basic Index are less systematic and may even be abbreviated or presented by molecular formula.

A systematic chemical substance name implies certain structural features and functional groups that individually may be important in a search. Each chemical substance word within a complete name that is separated by spacing or punctuation is indexed as a complete word. In addition, the special segmentation algorithm applied by DIALOG to other chemical information databases has been applied to all chemical names in the Basic Index of the Beilstein file. This chemical segmentation provides access to the chemically significant segments that represent structural features and functional groups

Table I. Data Fields Included in the Basic Index

Field Name	Search Suffix	Notes
INDIVIDUAL FIELDS		
Azeotrope Component	/AZ	in English
Biological Function	/BF	in German or English
Chemical Name	/CN	in English
Characterization Derivative	/DR	in German or English; complete phrase also in DR=
Ecological Data	/ED	in German or English
Element Count	/EC	
German Chemical Name	/GC	in German
Isolation from Natural Products	/IS	in German or English
Reagent	/REAGENT	in English; complete phrase also in REAGENT=
Starting Material	/START	in English; complete phrase also in START=; search also BN=
Synonym	/SY	in German or English; complete phrase also in SY=
Toxicity	/TX	in German or English
Use	/US	in German or English
Keywords of Short File	none	in English
Reference Tags	none	in English; complete phrase also in RT=
Other text not in a specific field	none	in German or English
COLLECTED FIELDS		
Comment (Other Conditions)	/COMMENT	in German or English; all comments attached to reported properties, preparations, or chemical reactions
Chemical Reactions	/CR	includes Partner, Product, Comment Temperature and Pressure
Chemical Name and Synonyms	/NA	in German or English; includes Chemical Name, German Chemical Name and Synonym; complete phrase also in NA=
Partner	/PARTNER	in English; includes partner in Chemical Reactions and partner in various multi-component systems such as Liquid/Liquid Systems, Liquid/Solid Systems, or Transport Phenomena; complete phrase also in PARTNER=
Preparation	/PR	includes Isolation from Natural Product, Starting Material, Reagent, (by-)Product, Other Conditions, Temperature and Pressure
Product	/PRODUCT	in English; includes productsin Chemical Reactions and by-products in Preparations
Solvent	/SOLVENT	in English; includes all solvents attached to reported properties; complete phrase also in SOLVENT=

embedded within a chemical name. For example, the term dichlorodibromomethane has been segmented into the following:

di
chloro
di
bromo
methane
bromomethane
dibromomethane
chlorodibromomethane

Therefore, in addition to searching a chemical name as a full name, the segments can be selected individually or with proximity operators to retrieve all complete names that contain the specified segments. Searching a single segment retrieves all words that contain the term either as a full word or as a segment. To select records where a segment occurs only as a separate word, the /FW suffix should be added to the select statement. The difference in retrieval between a segment and the same segment as a full word is illustrated in Figure 1.

```
? Select methane
      S1    17367   METHANE   (see also methan, methans)

? Type s1/bn,k

 1/BN,K/1
3120632
Crystals
  Melting Point: 166 - 171 C; Solvent:  nitromethane ; (Ref. 1)

? Select methane/fw
      S2     3408   METHANE/FW   (see also methan, methans)

? Type s2/bn,k

 2/BN,K/1
2707901
  Synonym:  Bis(2-methoxy-5-carboxyphenyl)methane
```

Figure 1. Comparison of Segment and Full Word retrieval.

Some segments are not only full words, but also complete names of substances of records. The easiest way to find the Beilstein record for one of these is to select the name with the NA= prefix, as illustrated in Figure 2. Two or more records may be retrieved because names and synonyms used by *Beilstein* do not indicate isotopic substitution or stereochemistry. These are, however, shown by the graphic structure. In this example, the record with BN=1813943 is for methane with the 12C isotope. The record with BN=1718732 is for methane with the naturally-occurring carbon isotope distribution.

Related Terms. One great advantage of Beilstein on DIALOG is that virtually every term in the database is searchable. As summarized in Table

I, Basic Index terms are a mixture of German words and English words with American or British spellings. Abbreviations and linear molecular formulas may also be present. DIALOG has incorporated the observed variations as scope notes and related terms to alert the searcher to consider these as additional search terms. For example, if a term such as water is selected, the result is as illustrated in Figure 3. Here the related terms are all listed in the scope note, but if the list of terms is too long, the scope note shows the number of related terms present instead of the terms themselves. The related terms for water and the number of records in which they occur are illustrated in Figure 3.

Related terms significantly enhance retrieval. Any non-relevant items retrieved by the broad search approach are usually eliminated by adding other search criteria. In Figure 3 retrieval has been restricted to the Comment field to illustrate the presence of both the German and molecular formula equivalents of the original search term water.

Keywords and Reference Tags. Non-numeric factual *Handbook* and Short File data is entered in the form of controlled vocabulary keywords. In addition, Short File specific keywords are used to indicate the existence of data that is too new to have been included in *Handbook* records. Keywords are searchable in the Basic Index as complete phrases and by individual words in the phrases. Complete phrases are also searchable in the Reference Tag, RT=, index discussed below.

Additional Indexes. The Additional Indexes in BEILSTEIN ONLINE provide access to most of the subject-oriented fields listed in Table I plus the physical property fields listing numerical data and keyword descriptions of properties. There are approximately 80 numeric fields and 30 phrase indexed fields used to present basic information about the substance, references and data present, preparation and isolation data, chemical reaction data, structure and energy parameters, physical state, physical and mechanical properties, transport phenomena, calorific data, optical properties, spectral data, magnetic properties, electrical properties, electrochemical behavior and multi-component systems. All fields in the Additional Indexes are searched using prefix codes. Some important fields and features are discussed below.

Data Present. The Data Present (DP=) field, in code form, and the corresponding Data Present Name (DPN=), in phrase form, are useful for determining the presence of a particular type of data. For example, to find all substances for which solubility data is given in the database, the search statement **S DP=SL** (or **S DPN=SOLUBILITY**) can be used.

Related Terms. To assist the searcher in determining the controlled vocabulary keywords that can be used to search non-numeric properties, the keywords available as Reference Tags have been incorporated as related terms with the appropriate Data Present code. Figure 4 illustrates how to find and use the Reference Tags associated with a specific data present. The desired Reference Tag(s) can then be selected directly using the appropriate R number(s) or the terms may be searched in the Basic Index.

Collected Data Fields. Many properties are reported with one or more reference parameters such as solvent, temperature, pressure or wavelength. To minimize the number of different search prefixes, they have been indexed in collected fields. These fields are summarized in Table II.

```
? s na=methane
      S3        2  NA=METHANE

? t s3/bi/1-2

 1/BI/1
1813943
  Lawson No: 9
  Beilstein Cit: 5-01
  Molecular Formula: CH4         [12C isotope]
  Molecular Weight: 16.04
  Synonym: Methane
  No. of Ref: 8
  Data Present:
 Data  Ref
 +Ref Only     UDF    Data Type
         1     CR  Chemical Reactions
         3     PP  Physical Properties
         2     KW  Short File Keywords

 1/BI/2
1718732        CAS Reg. No: 74-82-8 15802-48-9 20741-88-2 27680-53-1
                            59862-12-3 67969-37-3 69112-73-8
methane
  German Chem. Name: Methan
  Lawson No: 9
  Beilstein Cit: 5-01; 0-01-00-00056; 1-01-00-00004; 2-01-00-
      00003; 3-01-00-00001; 4-01-00-00003
  Molecular Formula: CH4
  Linear search formula: C1H4
  Molecular Weight: 16.04
  No. of Ref: 4645
 Data  Ref
 +Ref Only     UDF    Data Type
  197   12     PR  Preparative Data
 1114  780     CR  Chemical Reactions
  478  706     PP  Physical Properties
    1          PB  Physiological Data
    4          DR  Characterization Derivative
        47     KW  Short File Keywords
```

Figure 2. Finding a substance of Record with a Single-Word Name.

```
? s water
      S1        3156   WATER (see also H2O, w, wasser, wassers)

? e (water)

Ref     Items Type  RT  Index-term
R1      31566        4 *WATER (see also H2O, w, wasser, wassers)
R2      82025   R    4  H2O (see also w, wasser, wassers, water)
R3       1129   R    5  W (see also 5 related terms)
R4       5554   R    4  WASSER (see also H2O, w, wassers, water)
R5       1287   R    5  WASSERS (see also H2O, w, wasser, water,
                        waters)

? s r1:r5
      S6      110661  R1:R5

? s s6/comment
      S7       15404  S6/COMMENT

? t 7/bn,cn,k/21

 7/BN,CN,K/21
1813365
N, N'-bis-(2,2,2-trichloro-1-hydroxy-ethyl)-thiourea
Preparative Data
   Preparation
      Starting Material: anhydrodichloralthiourea
      Reagent: PCl5, POCl3
      Other Conditions: und nachfolgende Behandlung des
         Reaktionsproduktes mit Wasser (Ref. 1 handbook...

Crystals

  Melting Point: 79 - 80 C; Solvent: acetic acid, H2O ; (Comment:
      with: 0.5 Mol.  H2O (solvent).) (Ref. 1 handbook...
```

Figure 3. Related Terms Enhance Retrieval.

```
? e dpn=crystal phase

Ref      Items    Index-term
E1         109    DPN=CRITICAL VOLUME COMMENTS/REFS. (DP=CRVOLC)
E2           0    DPN=CROSS FILE REFERENCE (DP=CFR)
E3       10661   *DPN=CRYSTAL PHASE (DP=RTCP)
E4       10661    DPN=CRYSTAL PHASE COMMENTS/REFS. (DP=RTCPC)
E5        1123    DPN=CRYSTAL SYSTEM (DP=CSYS)
E6        1702    DPN=CRYSTAL SYSTEM COMMENTS/REFS. (DP=CSYSC)
E7           0    DPN=CRYSTALLIZATION MOLS OF SOLVENT (DP=CSOLMOL)
E8           0    DPN=CRYSTALLIZATION SOLVENT (DP=CSOL)
E9           0    DPN=CRYSTALLIZATION SOLVENT COMMENTS/REFS. (DP=CSOLC)
E10          0    DPN=DECOMPOSITION (DP=DECOMP)
E11       4776    DPN=DECOMPOSITION COMMENTS/REFS. (DP=DECOMPC)
E12          0    DPN=DECOMPOSITION SOLVENT (DP=DECOMPS)

? e (DP=RTCP)
Ref      Items  Type   RT  Index-term
R1       10661          9  *DP=RTCP (Crystal Phase)
R2         115    N     1  RT=ASSOCIATION IN THE SOLID STATE
R3          87    N     1  RT=CRYSTAL GROWTH
R4         246    N     1  RT=CRYSTAL HABIT
R5         506    N     1  RT=CRYSTAL MORPHOLOGY
R6        7814    N     1  RT=CRYSTAL STRUCTURE DETERMINATION
R7        1433    N     1  RT=INTERPLANAR SPACING
R8        1190    N     1  RT=POLYMORPHISM
R9          96    N     1  RT=RATE OF CRYSTALLIZATION
R10         58    N     1  RT=RATE OF TRANSITION

? s r6
         S1         7814  RT="CRYSTAL STRUCTURE DETERMINATION"

? t s1/id,pp
 1/ID,PP/1
3124217
  Molecular Formula: C21H20Br2O8
  Molecular Weight: 560.19
  Synonym: 2 alpha,7a alpha,3 beta,3a
          beta-Bis-(1,2-dimethoxycarbonyl-etheno)-4 alpha,7
          beta-dibrom-3a,4,7,7a-tetrahydroindan
  No. of Ref: 2
  Data Present:
Data   Ref
 +Ref  Only     UDF   Data Type
          1     PR  Preparative Data
   1      1     PP  Physical Properties
Crystals
   Melting Point: 182 - 183 C; (Ref. 2)
   Refs.
      2, Muir, Sim, JCSPAC, J.Chem.Soc.B, (1968)667
   Crystal Phase
      Crystal structure determination (Ref. 2)
      Refs.
         2, Muir, Sim, JCSPAC, J.Chem.Soc.B, (1968)667
```

Figure 4. Finding and using Reference Tags to search non-numeric data.

Table II. Collected Data Fields for Reference Parameters

Field	Search Prefix	Exceptions and Notes
Pressure	PRES=	Except Boiling Point Pressure (BPP=) and Sublimation Point Pressure (SUBPRES=).
Solvent	SOLVENT=	Also searchable by segment, word or phrase in the Basic Index or with the /SOLVENT suffix.
Temperature	TEMP=	Except Vapor Pressure Temperature (VPT=).
Wavelength	WL=	

There are two ways to take advantage of these collected fields. One is to use a collected field with the data present (code or name) to determine if the data is reported with the specified parameter. For example, to find all substances for which solubility in ethanol is reported, the (F) operator is used to link the data present with the specified solvent, as illustrated in Figure 5. The result of this type of search can be further modified by other search criteria such as other data present terms or substructure.

```
? s dpn=solubility (f) solvent=ethanol

          2009  DPN=SOLUBILITY  (DP=SL)
        221003  SOLVENT=ETHANOL
     S1    417  DPN=SOLUBILITY (F) SOLVENT=ETHANOL

? t 1/id,sl/1

 1/ID,SL/1
1812969
2-formyl-3-thioureido-acrylic acid ethyl ester
  German Chem. Name: 2-Formyl-3-thioureido-acrylsaeure-aethylester
  Molecular Formula: C7H10N2O3S
  Molecular Weight: 202.23
  No. of Ref: 1
  Data Present:
Data  Ref
 +Ref Only     UDF    Data Type
   1           PR   Preparative Data
   3           CR   Chemical Reactions
   2           PP   Physical Properties
Solution Behavior
   Solubility :  20 g/l; In Solution or in Pure Solvent Solvent:
     ethanol;(Ref. 1,  handbook)
   Solubility :  66.666 g/l; In Solution or in Pure Solvent
     Solvent: ethanol ; (Comment: in boiling solvent.) (Ref. 1,
       handbook)
   Refs.
      1, Dyer, Johnson, JACSAT, J.Amer.Chem.Soc., 56 (1934) 224
```

Figure 5. The (F) operator links the data present to a reference parameter.

The second approach is to link a collected field with a specific value or range of values for the associated property. For example, to find all substances that have a solubility of 30 - 35 grams/liter in water, the (S) operator is used to link the solubility to the solvent, as illustrated in Figure 6.

```
? s sl=30:35 (s) solvent=(water or h2o)
                  87  SL=30 : SL=35
                   0  SOLVENT=WATER
               70506  SOLVENT=H2O
       S3         32  SL=30:35 (S) SOLVENT=(WATER OR H2O)

? t s3/id,sl

 3/ID,SL/1
1811870        CAS Reg. No: 140-95-4
N, N'-bis-hydroxymethyl-urea
  German Chem. Name: N, N'-Bis-hydroxymethyl-harnstoff
  Molecular Formula: C3H8N2O3
  Molecular Weight: 120.11
  No. of Ref: 59
  Data Present:
 Data  Ref
 +Ref  Only      UDF     Data Type
    8     1      PR    Preparative Data
   56     8      CR    Chemical Reactions
   23     3      PP    Physical Properties
          4      KW    Short File Keywords
Solution Behavior
   Solubility:  32.29 g/l; In Solution or in Pure Solvent   Temp: 28 C;
        Solvent:  H2O ; (Ref. 27,  handbook)
   Solubility: 148 g/l; In Solution or in Pure Solvent   Temp: 18C;
        Solvent: H2O; (Ref. 21,  handbook)
Refs.
   21, Reichherzer, Chwala, Oest. Chemiker-Ztg., 51(1950)161, 162
   27, Klein, Tauboeck, BIZEA2, Biochem.Z., 241(1931)425
```

Figure 6. The (S) operator links a specific property value to a parameter.

Molecular Formulas. A molecular formula is given for each substance of record. The Molecular formula is in Hill Order, that is, carbon followed by hydrogen followed by all other elements in alphabetical order by element symbol. Molecular formulas are searched with the MF= prefix.

Several valuable search terms are derived from the main molecular formula of a record to permit generic searching based on the elements present in the substance. Table III summarizes these fields and their search prefixes. Each of these fields will be discussed briefly.

The Element Count is a valuable search term for specifying that an element can occur a certain number of times in a formula. The Element count is especially useful when it is not desirable to completely specify the composition of a substance. For example, to find all substances that contain fourteen carbons and one chlorine, the search statement **S EC=(C0014 AND**

numerical fields. Most of the numerical data fields are available as Handbook data. Five of the numerical data fields, Boiling Point (BP=), Melting Point (MP=), Density (DN=), Optical Rotatory Power (ORP=), and Refractive Index (RI=), are available as both *Handbook* data and Short File data.

Numerical values can be range searched by specifying the lower and upper limits with a colon between the two numbers. For example, to find substances with a reported boiling point between 100 and 110 degrees Celsius, the following search statement can be used, **S BP=100:110.**

The standard DIALOG relational operators can also be used to specify ranges of values. For example, to find substances with a refractive index greater than or equal to 1.45, the following search statement can be used, **S RI=>1.45.**

Figure 7 illustrates a search for substances that have a boiling point of 100 to 102 degrees Celsius at a pressure of 0.1 to 0.2 torr and have a refractive index greater than or equal to 1.55.

```
? s bp=100:102 (s) bpp=.1:.2 and ri=>1.55
            21035  BP=100 : BP=102
            59513  BPP=.1 : BPP=.2
             2396  (BP)(S)(BPP)
            38576  RI=>1.55
       S1     162  BP=100:102(S)BPP=.1:.2 AND RI=>1.55

? t s1/bi,bp,ri/1

 1/BI,BP,RI/1
3083182
  Lawson No: 7343
  Beilstein Cit: 5-07
  Molecular Formula: C14H16O
  Molecular Weight: 200.28
  Synonym: trans-1.2.3.4.4a.9.10.10a-Oktahydro-phenanthron-(10)
  No. of Ref: 1
  Data Present:
 Data  Ref
 +Ref Only      UDF     Data Type
         1      PR   Preparative Data
    2           PP   Physical Properties
         2      KW   Short File Keywords
 Liquids
   Boiling Point:  100 - 105 C;
 Pressure:  0.1 - 0.15 Torr; (Ref. 1)
   Refs.
      1, Parham, Czuba, JACSAT, J.Amer.Chem.Soc.,
         90(1968)4030,4037 Optical Properties
   Refractive Index:  1.5722; Wavelength: 589 nm; Temp: 27 C; (Ref.1)
   Refs.
      1, Parham, Czuba, JACSAT, J.Amer.Chem.Soc.,
         90(1968)4030,4037
```

Figure 7. (S) Proximity, range search, and relational operators are used to search numerical fields.

Table III. Data Fields Based on Molecular Formula

Field Name	Search Prefix	Notes and Examples
Molecular Formula	MF=	Hill order, e.g., MF=C14H22ClN3O3Si
Molecular Weight	MW=	g/mol, e.g., MW=343.88; use any numeric search
Element Count	EC=	S EC=C0014 AND EC=Cl0001; also range search, e.g., S EC=C0014:C0016
Molecular Elements	ME=	Hill order; all elements in the complete formula, e.g., ME=CHCLNOSI
Periodic Group Number	GN=	for all elements except C and H, e.g., GN=A4 for Si, GN=A5 for N, GN=A5 for O, and GN=A7 for Cl
Periodic Index Term	PI=	collection of group numbers, e.g., PI=A4567
Periodic Table Row	PT=	transition rows; e.g., PT=T1 if a 1st transition row element is present

Organometallic compounds and metal complexes of the transition elements are indexed by Group Number, as discussed above, and by the transition row number in the Periodic Transition Row (PT=) index. A formula such as C3H6Cl2Pt is retrieved by selecting PT=T3, GN=A7, GN=B8, or PI=A7B8.

Numerical Data Fields. As shown earlier, there are about 80 numerical data fields. These fields present the specific numerical values of the properties and are searchable using the standard DIALOG search techniques for **CL0001)** can be used. Ranges of element counts can also be searched. To find all substances with fourteen to sixteen carbon atoms, the search statement **S EC=C0014:C0016** can be used.

The Molecular Element (ME=) index lists the elements of a molecular formula by molecular symbol in Hill order omitting the number of occurrences of each element. For example, it is possible to find all fluorocarbons in the database by simply entering **S ME=CF.**

In addition to the above fields, molecular formulas are also indexed using terms that relate the elements in them to the periodic classification of elements outlined in a periodic chart. Each element in the formula, except carbon and hydrogen, is indexed by the Group Number (GN=) corresponding to the column of the periodic chart in which it occurs. A substance containing carbon, hydrogen, nitrogen and oxygen is indexed under GN=A5 and GN=A6. A search by group number specifies the group from which elements must be present; elements from other groups may or may not be present.

All group numbers associated with a substance are collected in the Periodic Index Term (PI=) index to more completely describe a substance, e.g., **S PI=A56** retrieves substances containing carbon, possibly hydrogen, any elements from group A5 (nitrogen, phosphorus, arsenic, antimony or bismuth) and any elements from group A6 (oxygen, sulfur, selenium, tellurium or polonium), but no other elements.

Lawson Number and Lawson Number Set. The value of the Lawson Number, a number generated by an algorithmic analysis of the structure, for searching related positional isomers has been published elsewhere *(1-3)*. Each substance record contains one or more Lawson Numbers for each of the chemically significant fragments in the structure. Each individual Lawson number assigned to a substance is searchable in the Lawson Number (LN=) index, e.g. **S LN=26333** retrieves all substances indexed with this number. The set of Lawson numbers assigned to the substance is searchable in the Lawson Number Set (LNSET=) index, e.g., **S LNSET=289 26333** retrieves all substances indexed with these two numbers and no others.

The LNs or the entire LNSET in a substance record can be saved for further searching using the MAP command. The MAP command creates a saved search strategy based on the terms in a specified field. The created search can be used immediately, or later, to search for addition records with the same terms. Figure 8 illustrates using a MAP LNSET to create a temporary search that is immediately executed to retrieve other similar substances. For more precise search results, the set created by the LNSET search can be further qualified by an element count search.

Special Data Fields. DIALOG has created some special searchable and displayable fields that give the user a better indication of the information in a given record. These fields are summarized in Table IV. All special data fields, except the Subfile field, are numeric and can be searched using the techniques discussed in the Numerical Data Fields section above.

Table IV. Special Fields Created by DIALOG

Field Name	Search Prefix	Notes
Number of Attributes	NAT=	The total number of different data fields present in the record
Number of Chemical Reactions	NCR=	The total number of chemical reactions with details or reference only in the record
Number of Lawson Numbers	NLN=	The total number of Lawson numbers in the record
Number of Preparations	NPR=	The total number of preparations with details or reference only in the record
Number of References	NRF=	Total number of cited references in the record
Subfile	SF=	Available terms are: SF=HANDBOOK, SF=SHORT, and SF=HANDBOOK SHORT. They provide a way to differentiate records with *Handbook* data or Short file data or both; at the record level

```
? s bn=128656
      S2        1  BN=128656

? t s2/bi

 2/BI/1
128656       CAS Reg. No: 26156-48-9
2-bromo-isonicotinic acid methyl ester
  German Chem. Name: 2-Brom-isonicotinsaeure-methylester
  Lawson No: 289, 26333
  Beilstein Cit: 5-22; 4-22-00-00685
  Molecular Formula: C7H6BrNO2
  Molecular Weight: 216.03
  No. of Ref: 4
  Data Present:
 Data  Ref
 +Ref Only      UDF     Data Type
    1    1      PR   Preparative Data
    1           CR   Chemical Reactions
    2           PP   Physical Properties
         1      KW   Short File Keywords

? map lnset t exs s2

1 select statement(s)
serial#TC160

Executing TC160
      S3       44  LNSET=289 26333

? t s3/bn,na,ln/1,10,20,40

 3/BN,NA,LN/1
477026
  Lawson No: 289, 26333
  Synonym: 5-Nitropicolinsaeuremethylester

 3/BN,NA,LN/10
472356
  Lawson No: 289, 26333
  Synonym: Methyl-bromo-3-pyridincarboxylat-4

 3/BN,NA,LN/20
172043
6-nitro-nicotinic acid methyl ester
  German Chem. Name: 6-Nitro-nicotinsaeure-methylester
  Lawson No: 289, 26333

 3/BN,NA,LN/40
128572
6-fluoro-pyridine-2-carboxylic acid methyl ester
  German Chem. Name: 6-Fluor-pyridin-2-carbonsaeure-methylester
  Lawson No: 289, 26333
```

Figure 8. MAP LNSET is used to find similar substances.

CHEMICAL REACTION SEARCHING

Chemical Reactions in Preparations and Chemical Reactions Fields. The *Beilstein Handbook* is a rich source of chemical reaction data. Until the availability of BEILSTEIN ONLINE, most of this data was not easily searched. On DIALOG these chemical reaction data may be searched comprehensively. Chemical reaction data are found in the following fields: Preparations, Isolations from Natural Products, Chemical Reactions and Derivatives. Approximately 90% of the records in the *Handbook* contain preparations including isolations from natural products and about 15% of the records contain chemical reaction data. Therefore, preparations are important sources of chemical reaction data.

Figures 9 and 10 outline the reaction schemes for preparations and chemical reactions. Note that reactions may be searched by the component fields of the reaction and reaction conditions. Also, reactions can be searched without specifying the reaction component using /CR or /PR depending on the

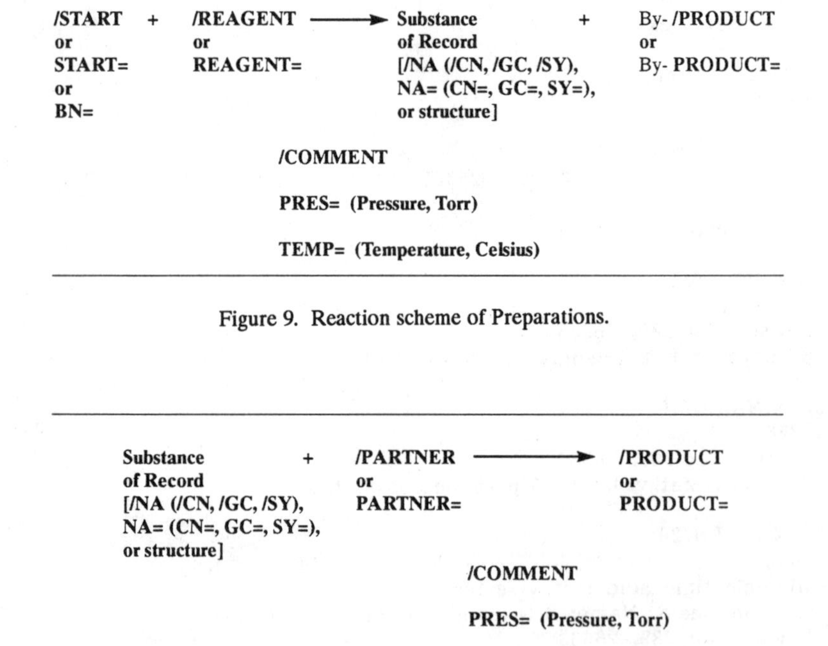

Figure 9. Reaction scheme of Preparations.

Figure 10. Reaction scheme of Chemical Reactions.

source of the data. This is important because of the way the *Beilstein Handbook* database was compiled. The *Handbook* data, which contains references back to 1779, were not gathered originally for an online database. Indexers have done their best to assign the data to various fields. For example, in preparations, catalysts may be listed as reagents or entered as part of the text string in the other conditions field. In Figure 11, a search for Lindlar catalyst results in 168 records containing Lindlar as a reagent, however, 33 more records are recovered searching Lindlar in the entire preparation field.

```
? s lindlar/reagent
      S1      168 LINDLAR/REAGENT

? s lindlar/pr
      S2      201 LINDLAR/PR

? s s2 not s1
              201 S2
              168 S1
      S3       33 S2 NOT S1
? t 1/k/1,15

  1/K/1
Preparative Data
 Preparation
  Starting Material:BN=1762545, 3-allyloxy-3,7-dimethyl-oct-6-en-1-yne
  Reagent:  Lindlar -catalyst, hexane
  Other Conditions: Hydrogenation (Ref. 1,  handbook ? t 2/k/15

  2/K/15
Preparative Data
   Preparation
      Starting Material: BN=1785141, 3-methyl-eicos-1-yn-3-ol
      Other Conditions: bei der Hydrierung an Lindlar -Katalysator (Ref. 1,
         handbook...
```

Figure 11. Searching a catalyst anywhere in the Preparation field enhances retrieval.

Comparison of Searchable Fields for Preparations and Chemical Reactions on DIALOG vs STN. Because of the variable indexing of the reagents, partners, products etc. for chemical reactions and preparations, DIALOG has made virtually every word searchable in these fields. Tables V and VI compare the BEILSTEIN search fields for Preparations and Chemical Reactions on DIALOG with those in BEILSTEIN on STN International. Furthermore, in these fields, all systematic chemical names detected are subject to the DIALOG segmentation process and may be searched using chemically significant segments.

Table V. Searchable Fields for Preparations

Field Name	DIALOG	STN
Starting Material (Educt)	yes	yes
Reagent	yes	yes
By-Product	yes	yes
Comments (Other Conditions)	yes	no
Temperature	yes	no
Pressure	yes	no
Number of Preparations	yes	no

Table VI. Searchable Fields for Chemical Reactions

Field Name	DIALOG	STN
Partner	yes	yes
Product	yes	yes
Subject (Aim of Study)	yes	yes
Temperature	yes	no
Pressure	yes	no
Comments (Other Conditions)	yes	no
Number of Reactions	yes	no

STRUCTURE SEARCHING

Graphic structure searching was introduced on DIALOG with the release of BEILSTEIN ONLINE. Use of the Query Structure (QS) command invokes the structure search system on DIALOG. This system is a combination of S4, a substructure search system produced by SOFTRON GmbH., and the Query Structure commands developed by DIALOG. S4, described in Chapter SSS of this monograph, *(4)* uses a unique data structure that processes structure searches faster and more efficiently than other structure search systems.

ROSDAL Description of a Structure. For searching on DIALOG, a chemical structure is translated to a ROSDAL (Representation of Organic Structure

Description Arranged Linearly) character string. Three software packages are available for translating a graphically entered structure to a ROSDAL string: MOLKICK from Softron GmbH *(5)*, ChemConnection from SoftShell International Ltd. *(6)*, and ChemTalk Plus from Molecular Design Limited. *(7)*. Also, a chemist can encode a structure using the publication, "Online Searching on DIALOG: Rosdal Strings, Structure Search via Keyboard Commands" available from DIALOG and Springer Verlag *(8)*. This publication may be used by graphic structure software publishers to upgrade programs that send or receive graphic structure data. Figure 12 illustrates the writing of a ROSDAL string to search for benzofuran with any halogen at the 3 position.

---

1. Draw the structure of the benzofuran ring and number the atoms in any sequence.

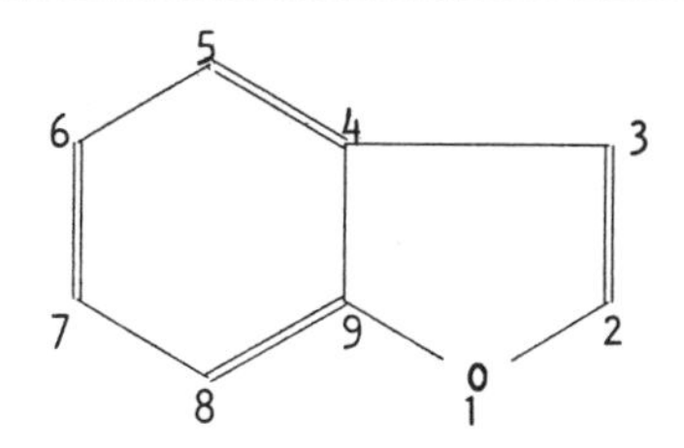

2. Write the ROSDAL string describing the bonding of the ring system using anofor single bonds and an = sign for double bonds. Use the element symbol immediately after the number to indicate a non-carbon atom.

1O -2=3-4=5-6=7-8=9-1, 2-7

Since atoms 2 through 9 are all Carbons, and the bonds in the sequence from atom 1 to atom 9 alternate between single and double, the ROSDAL string can be compressed using -= between the two nodes.

1O -= 9-1,2-7

3. Attach an X (indicating any halogen) to the number three atom in the ring, give it a number and add the bonding description to the already written ROSDAL string.

1O -=9-1, 2-7, 3-10X

4. Add ".@" to terminate the ROSDAL string: 1O -=9-1,2-7,3-10 .@

Figure 12. Summary of steps involved in writing a ROSDAL string.

Structure Search Commands. The Query Structure (QS) commands are a new set of commands on DIALOG. Table VII outlines the basic commands for this system. For a substructure search, a free test search on about 5% of the database may be performed to check the validity of the structure. Once validated, the test search may be run against the entire database by entering, QS FULL. DIALOG automatically searches the complete database using the last ROSDAL string entered.

After processing the query, DIALOG displays the results as a normal DIALOG set with no further selection necessary. This set can be displayed directly, combined with other DIALOG sets or used in other commands. Also, there is no maximum set size. For example, although it is unlikely that anyone would want to, a user could select a methyl group with anything attached. The resultant set includes about 75% of the current BEILSTEIN database, over 2 million records.

The user, at his option, can retrieve two-dimensional or stereochemical representations of the structures in a set. The GS display code is used to display a structure with stereochemical representation. The GR display code is used to display the structure without stereochemistry.

OUTPUT

Free Display Formats. Records in the BEILSTEIN file range from a single data item to over 250 pages of data. To assist the user in evaluating the data, we have provided two free formats that identify the substance of the record and indicate the approximate size of the record. Format 2 lists the types and amount of data in fields present in the record. All data listed after basic identity information on the substance contains references. A sample record of Format 2 is in Figure 13. Some of the datafields contain only references whereas other datafields in a record contain data and the corresponding references. Format 2 gives an indication of how much data with references are in the record.

The other free format is the ID format consisting of the number of preparations, chemical reactions, references and physical properties in a record.

KWIC and HILIGHT. KWIC, Keyword in Context, is a browsing format that allows the viewing of only those portions of the record that contain the selected search terms. Additionally, the hilight option is a feature which highlights the selected search terms in a record for rapid location of the terms. Both KWIC and HILIGHT are free. HILIGHT can be used with the KWIC format or any other output format for BEILSTEIN ONLINE records. See Figure 14 for an example of KWIC and HILIGHT.

Other Format Options. Because of the variation in the size of the records in this database, format options range from the complete record to sections of the record to specific parts of the record. The complete record may be displayed with the references at the end of the entire record or displayed after each section in which they appear. For large records, it is useful to have the references split by section for ease of viewing. Additionally, sections of the record, such as reaction data which includes preparations and chemical reactions may be displayed. An even smaller subset of the record may be displayed using a set of format codes that correspond to various physical properties, chemical name, molecular formula and other parts of the record.

```
 1/2/1
2349
2-amino-3H-pyrimidin-4-one
 German Chem. Name: 2-Amino-3H-pyrimidin-4-on
 CAS RN: 108-53-2, 70329-14-5
 Molecular Formula: C4H5N3O
 Molecular Wt: 111.1
 General Comment: and tautomeres.
 No. Ref: 41
 Data Present:
Data  Ref
+Ref Only    UDF    Data Type
        8    SD   Constitutional Data
        8         .Related Structure
   7    1    PR   Preparative Data
   7    1         .Preparation
   5    1    CR   Chemical Reactions
  23   14    PP   Physical Properties
        4    SE   .Structure & Energy Parameters
        3         ..Interatomic Distances & Angles
        1         ..Coupling Phenomena
  23   10         .Physical Properties of Pure Compound
  11    6    PS   ..Physical State
  11    6    PC   ...Crystals
   1              ....Color or Other Properties
   8         MP   ....Melting Point
        2         ....Crystal Phase
   1    1         ....Crystal System
        1         ....Space Group
        1         ....Dimension of Unit Cell
   1    1         ....Density of Crystal
  10    2    SP   ..Spectra
   1    1         ...NMR
        1         ....NMR Spectrum
   1              ....NMR Absorption
   1              ...Vibrational Spectra
   1         IR   ....IR Spectrum
   8    1         ...Electronic Spectra
   8    1         ....UV/VIS Spectrum
        1    EL   ..Electrical Properties
   2    1    EB   ..Electrochemical Behavior
   2    1         ...Dissociation Exponent
        6    KW   Short File Keywords
```

Figure 13. Data present (Format 2) output.

```
set hi *
HILIGHT set on as '*'
? s ethylenediamine/cn and nitrate/cr (s) hydrolysis/cr
              1884  ETHYLENEDIAMINE/CN  (see also aethylendiamin)
               901  NITRATE/CR  (see also nitrat)
              7302  HYDROLYSIS/CR  (see also hydrolyse)
                20  NITRATE/CR (S) HYDROLYSIS/CR
      S1         1  ETHYLENEDIAMINE/CN AND NITRATE/CR (S)
                        HYDROLYSIS/CR
? t 1/kwic/1

  1/KWIC/1
 N,N'-bis-(2)thienylmethylen-*ethylenediamine*
Chemical Reactions
   ...Chemical Reaction:
      Partner: aqueous ethanol, nickel (II)-*nitrate*
      Temp: 20 - 35 C
      Further Conditions: *Hydrolysis*
      Reaction Product: thiophene-2-carbaldehyde, ethylenediamine-metal
         complex
      Aim of the Study: Rate constant (Ref. 4,  handbook
   ...Chemical Reaction:
      Partner: aqueous ethanol, nickel(II)-*nitrate*
      Temp: 20 - 35 C
      Further Conditions: *Hydrolysis*
      Reaction Product: thiophene-2-carbaldehyde, ethylenediamine-metal
         complex
      Aim of the Study: Rate constant (Ref. 4,  handbook...
```

Figure 14. Keyword in Context (KWIC) and HILIGHT.

Structures. Structures for BEILSTEIN ONLINE are stored as ROSDAL strings. These ROSDAL strings are similar to the structure query ROSDAL except that x-y coordinates are also included. These x-y coordinates allow the plotting of the chemical structure. Software packages such as MOLKICK, ChemConnection, ChemTalk Plus and GEOFF, a PC software tool available from Dialog Information Services *(9)*, can be used to display the structures from the ROSDAL output from the database. GEOFF (Graphic Enhanced Output File Formater) is a low cost software tool that converts ROSDAL strings within a record output to a structure. It does not alter the textual information but simply replaces the ROSDAL string with a structure.

Stereochemical Structures. BEILSTEIN ONLINE contains structures with stereochemistry. The user has the option to display structures with two-dimensional orientation or with stereochemical orientation when the data is available. Figure 15 illustrates this.

CONCLUSION

The conversion of the *Beilstein Handbook* to the BEILSTEIN ONLINE database and the implementation of the database on DIALOG with new commands and features developed especially for the *Beilstein* database, have combined to give the user easy and varied access to a wealth of information on organic substances, their properties, structures and chemical reactions.

GS

1/GS/1
Graphic Structure:
'81689'

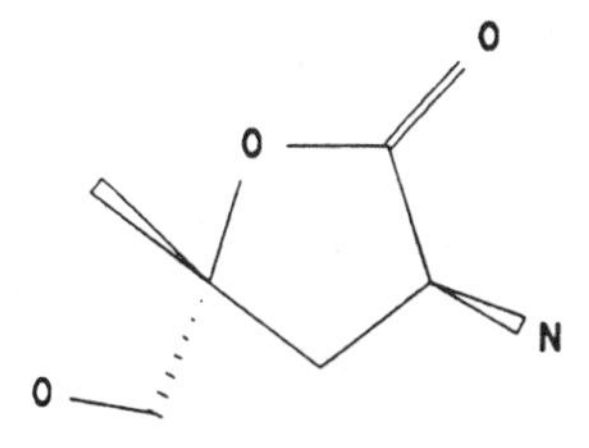

Display code GS shows structure, with stereochemistry when available.

GR

1/GR/1
Graphic Structure:
'81689'

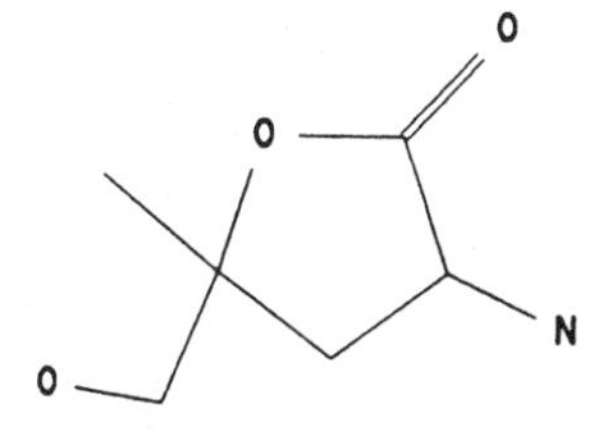

Display code GR shows structure without stereochemistry.

Structure graphics from GEOFF

Figure 15. Graphic structure display options.

REFERENCES

1. Lawson, A. J. In *Software-Entwicklung in der Chemie 2*, Gasteiger, J. Ed., Springer Verlag, Heidelberg, 1988, p. 1.
2. Lawson, A. J. In *Graphics for Chemical Structures*, Warr, W. A. Ed., ACS Symposium Series 341, American Chemical Society, Washington, DC, 1987, pp 80-87.
3. Lawson, A. J., In *The Beilstein Online Database*, ACS Symposium Series, American Chemical Society, Washington DC, 1989, Chapter 10.
4. Welford, S. M. In *The Beilstein Online Database*, ACS Symposium Series, American Chemical Society, Washington DC, 1989, Chapter 5.
5. MOLKICK, a product of Softron GmbH. (personal communication) Available from: Springer Verlag New York, Inc. 175 Fifth Avenue, New York, NY 10010 Springer Verlag Heidelberg New Media Department, Tiengartenstr. 17, D-6900 Heidelberg Germany
6. ChemConnection, a product of SoftShell International Ltd. (personal communication) Available from: SoftShell International, Ltd. 2754 Compass Drive Suite 375 Grand Jct., CO 81506
7. ChemTalk, part of the Chemist's Personal Software Series (personal communication) Available from: Molecular Design Limited 2132 Farallon Drive San Leandro, CA 94577
8. Welford, S. M., "Online Searching on DIALOG: Rosdal Strings, Structure Search via Keyboard Commands." Available from: Dialog Information Services, Inc. (address in #9) or Springer Verlag (address in #5).
9. GEOFF, Graphic Enhanced Output File Formatter, a PC software tool (personal communication) Available from: Dialog Information Services Marketing Department 3460 Hillview Avenue Palo Alto, CA 94304

RECEIVED May 17, 1990

Chapter 5

Chemical Structure Searching

Using S4/MOLKICK on DIALOG

Stephen M. Welford

Springer–Verlag, Tiergartenstrasse 17, Heidelberg, Federal Republic of Germany

The Beilstein Online database on DIALOG presents many exciting new possibilities for structure and data searching of organic chemical compounds. Structure searching on DIALOG is carried out using the Softron S4 structure search system, and structure queries can be built and submitted from your PC using the graphics-based query editor MOLKICK, developed jointly by Softron GmbH and the Beilstein Institute, and distributed by Springer-Verlag. This chapter describes the major features of the Beilstein structure file, the S4 search system implemented on DIALOG, and the use of MOLKICK for building and submitting structure queries.

Previous chapters of this book have described the Beilstein Handbook of Organic Chemistry and the development of the Beilstein Online database. The different implementations of the Beilstein database on two major hosts, STN International and DIALOG, have also been described. By the time you read this book, Beilstein Online will be being used successfully by many hundreds of organisations and chemists world-wide. This is an exciting time at which to explore the content of this important database which powerfully complements the existing online structure and bibliographic databases.

Structure searching in the Beilstein Online database is carried out with a new and powerful software system, S4. This system has been developed by Softron GmbH, in close cooperation with the Beilstein Institute and with DIALOG, and differs markedly in architecture and performance from existing structure search systems. In addition, DIALOG users will have available to them the flexible PC graphics-based query editor program MOLKICK, also produced by Softron GmbH and distributed by Springer-Verlag, which is tailored precisely to the S4 search system. As new capabilities in S4 are introduced, these will be reflected immediately in new versions of MOLKICK.

0097–6156/90/0436–0064$06.00/0

This chapter introduces you to these component parts of structure searching on DIALOG. We first review aspects of the Beilstein structure file and its characteristics on DIALOG, then describe briefly the S4 search system and how it relates to other existing systems for searching large structure files. Since PC-host communications are often a source of confusion, especially where graphics terminal emulation for the input and output of chemical structure diagrams is involved, we look closely at how MOLKICK is used to provide an interface to DIALOG. The use of MOLKICK for drawing and submitting query structures and displaying retrieved hits is described, and the capabilities of the S4 search system are explored by means of examples which illustrate some of the types of structure searching which are possible. Further examples of structure searching in the Beilstein database on DIALOG and on STN can be found in other chapters of this book.

The Beilstein Structure File

The Beilstein structure file contains organic chemical compounds which have been reported in the chemical literature since 1830. Production of the file from the printed Beilstein Handbook and from the literature abstracts at the Beilstein Institute is still in progress (1), so the size and content of the online file will continue to change until 1991, after which time the file will continue to grow steadily as the current chemical literature is abstracted. At the time of this Symposium the file contained 1.7 million compounds, and is expected to reach 3.5 million compounds in 1991.

This is smaller than the number of substances present in the Chemical Abstracts Service (CAS) Registry file (2). There are several reasons for this. Chemical Abstracts deals with all branches of chemistry, while Beilstein deals only with organic compounds of known constitution; organo-metallic compounds and inorganic substances are found in the Gmelin database, which will also soon be available online. Furthermore, Beilstein deals only with compounds for which measured and verifiable data are reported; this at least implies that the compounds have been isolated or observed in a pure state and are therefore worth recording.

A proportion of the compounds in the Beilstein structure file are compounds of known absolute or relative configuration; this is evident if you page through a volume of the Beilstein Handbook. Different stereoisomers of a compound are recorded separately in the Beilstein database, and often have markedly different physical and chemical properties, particularly optical properties and chemical reactivities, Figure 1. Accordingly, these have different Beilstein Registry Numbers (BRNs) in Beilstein Online.

2-[3-Methoxycarbonyl-1,3-dimethyl-2-oxo-cyclohexyl]-benzoesäure $C_{17}H_{20}O_5$.

a) 2-[(1*R*)-3*c*-Methoxycarbonyl-1,3*t*-dimethyl-2-oxo-cyclohex-*r*-yl]-benzoesäure, Formel XIII.

B. Aus Podocarpa-8,11,13-trien-15-säure-methylester mit Hilfe von CrO_3 (*Ohta, Ohmori*, Pharm. Bl. 5 [1957] 91, 95).

Kristalle (aus Me.); F: 151 – 152°. $[\alpha]_D^{25}$: –8,4° [A.; c = 2,4]. λ_{max} (A.): 253 nm und 295 nm.

b) 2-[(1*S*)-3*t*-Methoxycarbonyl-1,3*c*-dimethyl-2-oxo-cyclohex-*r*-yl]-benzoesäure, Formel XIV.

B. Analog dem unter a) beschriebenen Stereoisomeren (*Ohta, Ohmori*, Pharm. Bl. 5 [1957] 91, 95). Aus (4a*R*)-1*t*,4a-Dimethyl-9-oxo-(4a*r*,9a*c*)-1,2,3,4,4a,9a-hexahydro-fluoren-1*c*-carbon= säure-methylester beim Behandeln mit CrO_3 in Essigsäure (*Ohta*, Pharm. Bl. 5 [1957] 256, 259).

Kristalle; F: 128 – 131,5° [aus Me.] (*Ohta, Oh.*). 129 – 130° [aus Diisopropyläther; nach lang= samer Kristallisation] (*Ohta*) bzw. F: 104 – 106° [aus Diisopropyläther; nach schneller Kristalli= sation] (*Ohta*). $[\alpha]_D^{25}$: –35,4 [A.; c = 3]; λ_{max} (A.): 238 nm und 284 nm (*Ohta, Oh.*).

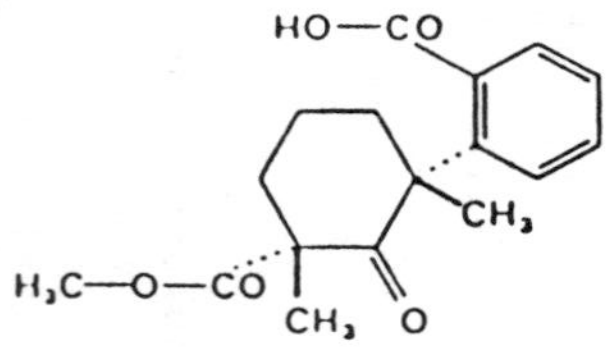

XIII 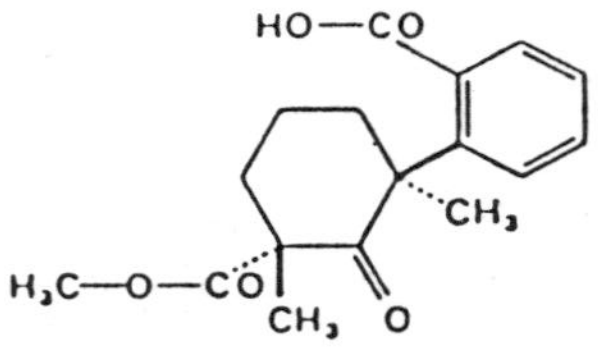   XIV

Figure 1. Stereoisomers with different physical properties (Reproduced from the Beilstein Handbook. Copyright 1984 Springer-Verlag).

On STN, all separately registered stereoisomers of a compound will be retrieved in a full structure search since the search is carried out on the basis of constitution only. When fully implemented on DIALOG, S4 will enable these stereoisomers to be searched selectively since the the configuration of the molecule is also taken into account by the search system. Stereospecific searching will be possible by indicating relative bond orientations and Cahn-Ingold-Prelog (CIP) configuration descriptors in the query structure.

Structural tautomerism is also a feature of the Beilstein structure file, and is common in many classes of organic compounds. Since analytical techniques are becoming increasingly sophisticated it is not uncommon to observe and characterise different tautomeric forms of a compound in the laboratory. Where these are reported in the literature, different tautomeric forms are recorded separately in the Beilstein database, each with a different BRN. In the STN implementation of the Beilstein database, tautomers are normalised according to the CAS Registry III conventions (3), and become represented by the same search record; consequently, different tautomeric forms of a compound are always retrieved collectively. In contrast, S4 on DIALOG will allow tautomeric forms to be searched selectively or collectively as desired.

The S4 Structure Search System

Searching large files of chemical structures places heavy demands on computer resources and, as a result, is a major component of the cost of online chemical information retrieval. The systems which are in use today are the result of much research and development aimed at maximizing the efficiency and flexibility of structure searching, and particularly substructure and generic structure searching, while minimizing the resources required and ultimately the cost met by the online searcher (4).

Since its first release in 1980 the CAS Registry file has been searched online by a system in which elaborate hardware is used to gain efficiency (5). When the STN implementation of the Beilstein database was made at FIZ Karlsruhe in 1988, enhanced software was developed which enabled the structure search system to be run on a single large mainframe computer (6). During this period, and in addition to the Telesystemes-DARC system used successfully with the CAS Registry file, two other structure search systems for large structure files have been developed independently - the HTSS system (7) which will be employed with the Beilstein structure file on Maxwell Online, and Softron's S4 system which is used on DIALOG. Both of these systems involve new and interesting datastructure and file designs, and deliver fast and efficient structure searching.

In the case of S4, a search file is built from the Beilstein structure file by encoding each atom of each molecule in the file in terms of its extended connectivity. From this large file of atom codes a search tree is generated, which forms the index through which the atom code file is accessed during a search. The most discriminating code which encompasses all relevant aspects of the query structure is used to search the index, and results in a list of addresses in the atom code file. The atom code file is then accessed and read sequentially from each starting address until the coding changes. Contained in this list are the hits. In most cases, no atom-by-atom search is necessary and, because of the minimized number of disk accesses in the sequential file, very fast search times can be achieved.

A benchmark of these various systems at their current stages of development has recently been carried out by the Beilstein Institute (Hicks, M. G., Jochum, C. J., Maier, H., Anal. Chim. Acta., in press). A further important characteristic of the structure search systems is the relationship between increasing file size and online search time. This relationship was investigated for S4 in the benchmark tests, and the average search time for S4 was found to increase at a very much smaller rate, ie. less than linear, than the increase in file size, Figure 2.

Online Communication with DIALOG

In order to perform structure searches in the Beilstein database you must be able to describe and communicate your query structure to DIALOG. MOLKICK is ideal for this purpose, and provides all the facilities you need for drawing, editing and saving your query structures. Before describing these functions of MOLKICK, we look briefly at the communication between your PC and DIALOG. PCs are now regularly used for communicating with online search services (8). To make this possible your PC must behave as if it were connected directly to the host computer; this is achieved by running on your PC a terminal emulation program, or more generally a communications package, and connecting the PC indirectly to the host computer via a telecommunications network by means of a modem. There are now many communications packages available, eg. DIALOG's DIALOG-LINK, PCPLOT, INFOLOG and CROSSTALK, most of which allow your PC to emulate a variety of different types of terminal, including graphics terminals running under several of the standard graphics protocols, eg. PLOT-10. In addition, the communications software controls the operation of the modem and provides automatic logon and data capture and upload facilities.

In addition to MOLKICK, several packages are now available which are designed specifically for chemical structure searching, eg. STN Express, DARC CHEMLINK, MDL

CHEMTALK, CHEMCONNECTION. These and others are reviewed in a recent ACS publication (9). These packages are designed for different purposes and, in most cases, are limited to use with only one particular host, eg. STN Express and DARC CHEMLINK. With the exception of MOLKICK these packages include the communications software which is necessary for the PC to communicate with the host. MOLKICK does not contain communications software, but is a smaller program designed to be used in conjunction with communications software which you may already have on your PC. MOLKICK is available as a terminate-stay-resident (TSR) program, which enables you to call up the MOLKICK structure editor at any time and from within any other program by typing the so-called "hot-key" combination. You can work with MOLKICK and then push it into the background, recalling it when you wish to continue drawing or editing a structure; the current status of your work in MOLKICK is retained at all times.

MOLKICK differs from other packages in another important respect. MOLKICK may be used for searching structure files on several different hosts; in addition to DIALOG, MOLKICK can be used for searching the Beilstein and CAS Registry files on STN and the CAS Registry file on Telesystemes-DARC. MOLKICK achieves this by converting your query structure into the form of structure description required by your chosen host. Your communications software is then used to upload this description to the host for searching.

We concentrate now on the use of MOLKICK with DIALOG, providing first a brief review of the structure drawing functions of MOLKICK.

The Query Structure Editor MOLKICK

MOLKICK provides all of the capabilities which you would expect in a state-of-the-art structure editor. Specifically, these include free-hand structure drawing using the mouse, extensive template libraries which you can build yourself, functions for resizing, moving, stretching and rotating on the screen groups and fragments of the structure, various structure display options, and an interactive filing system which allows you to store and retrieve upto 800 query structures. Atom and bond attributes and sites of allowed further substitution can be specified, and Markush-type generic structures can be built, incorporating a variety of generic groups such as "alkyl", "carbocyclic" and "heteroaryl".

Using MOLKICK is a highly interactive process; it is very difficult to provide the "feel" of such a program on the pages of this book. Nor is it possible here to describe all of the capabilities of MOLKICK. You will discover these for yourself when you sit at your PC with MOLKICK on the screen in front of you. We illustrate here the creation and editing of a query structure using

MOLKICK, and then describe the manner and form in which the query structure is transmitted to DIALOG for search using the S4 search system.

We are interested in investigating the properties of monoalkoxyisoquinolines, and have prepared in the laboratory 5-methoxyisoquinoline

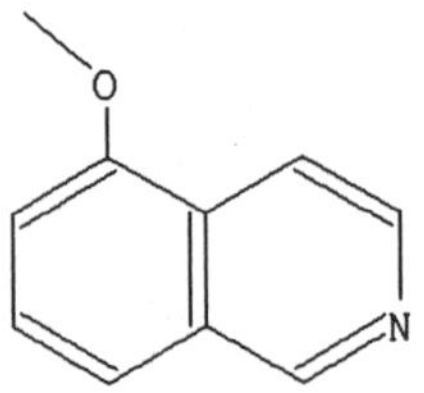

Study of this compound leads us to begin looking at the family of alkoxyisoquinolines in which the benzene ring is mono- or multi-substituted. In order to search for this class of compounds in the Beilstein database on DIALOG, we build the query structure using MOLKICK as follows.

From MOLKICK's edit window, we load the ring template library RING.DAT and select the naphthalene ring, Figure 3. The naphthalene ring appears in the edit window and can be moved to any part of the screen, Figure 4. The isoquinoline ring is created by changing the 2-carbon to a nitrogen atom in the atom window, Figure 5. In order to specify possible substitution on the benzene ring, a G-group is added to each ring atom, Figure 6. Selecting any G1 causes the structure to move to the upper-left portion of the edit window, leaving the remainder of the screen available for defining the possible values of G1, Figure 7. We define G1 as either a hydrogen or an alkoxy group; selecting the appropriate atom and pressing the function key F1 allows us to page through MOLKICK's context-sensitive help information to find which symbol should be used in conjunction with the oxygen atom to define the alkoxy group, Figure 8. To complete the query we must specify how the values of G1 are attached to the main structure. Selecting the oxygen atom of the alkoxy group and highlighting Attachments in the atom window opens the attachments window, in which we specify that the oxygen atom is attached to any atom in the main structure highlighted with the symbol 1, Figure 9.

During the creation of the query structure MOLKICK maintains a linear description of the structure in the form of a ROSDAL string (Representation of Organic Structure Diagrams Arranged Linearly). When you press the function key F7 on your keyboard the current ROSDAL string:

```
1=-4-14G1,1-11G1,1-7=-10N-5=-7,2-12G1,3-13G1,4-8;G1=(1H;2O
&-3ALK)
```

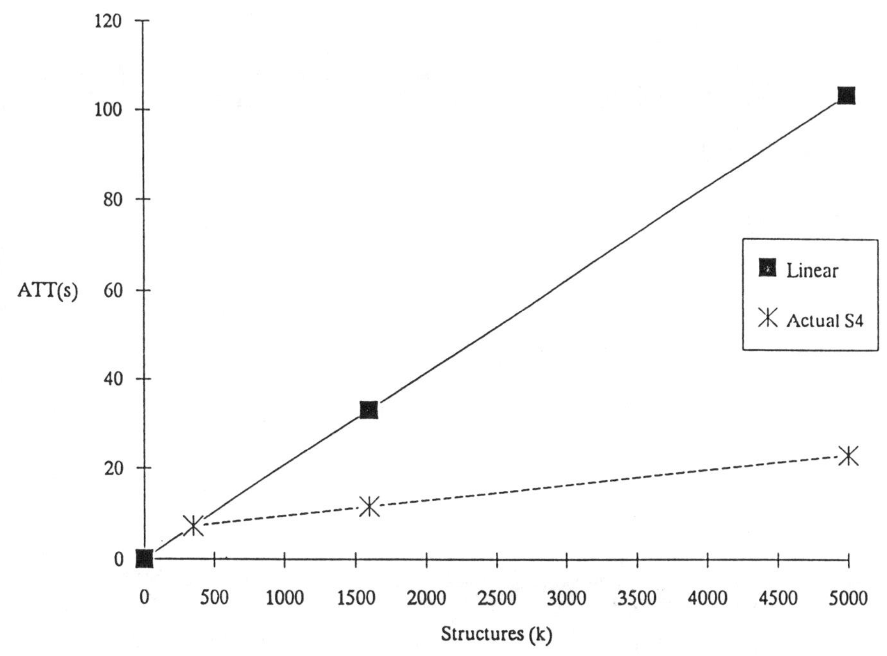

Figure 2. S4 search time increase with file size (Reproduced with permission of the Beilstein Institute).

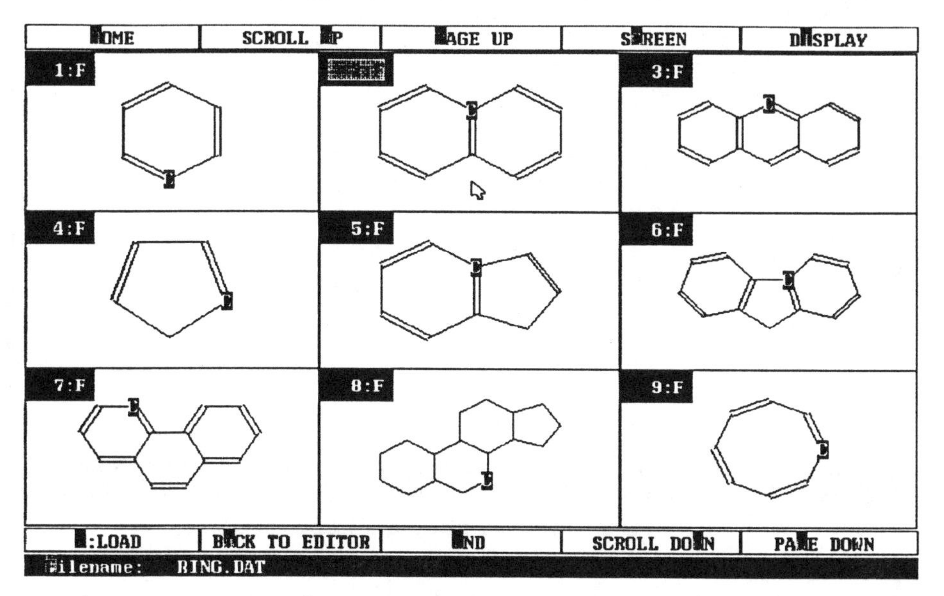

Figure 3. Selecting a structure from the ring file.

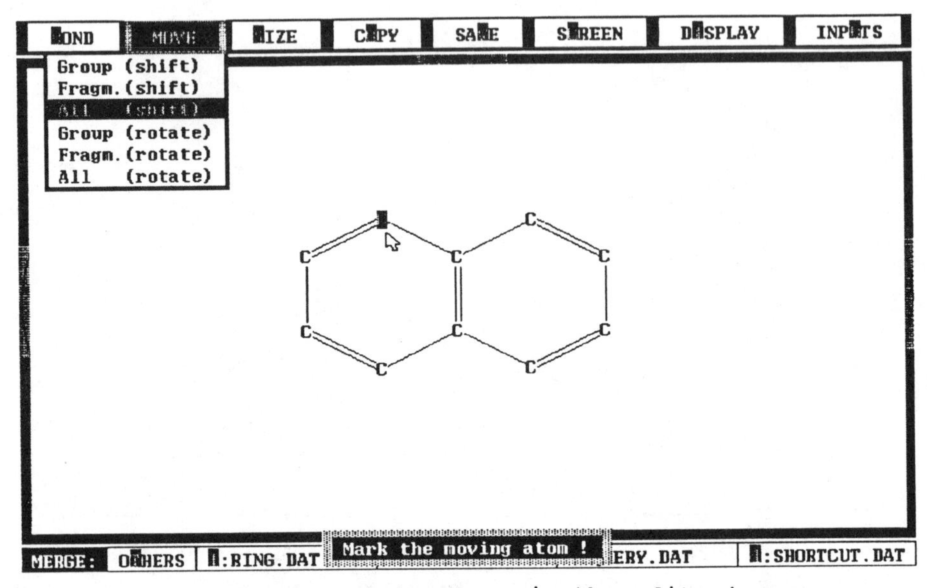

Figure 4. Moving a structure in the edit window.

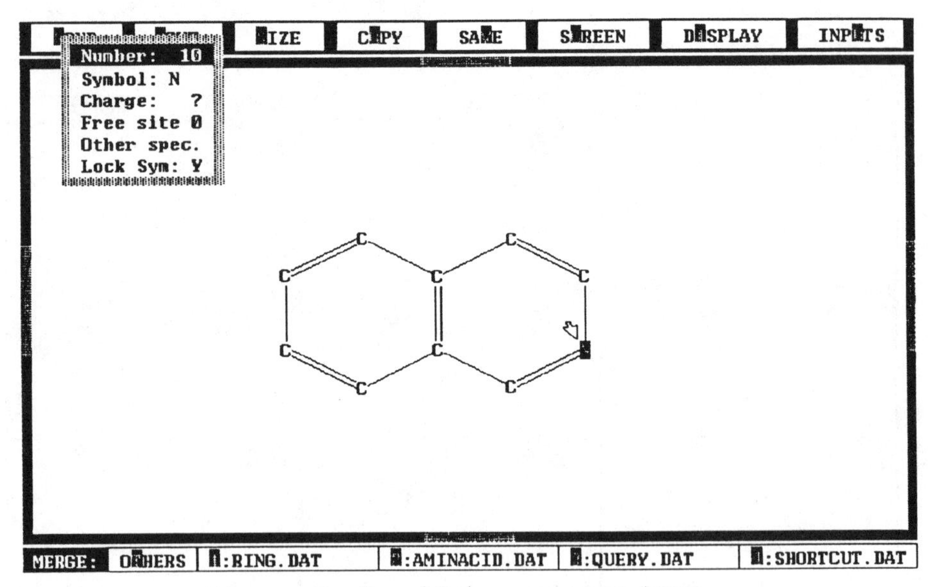

Figure 5. Specifying an atom type.

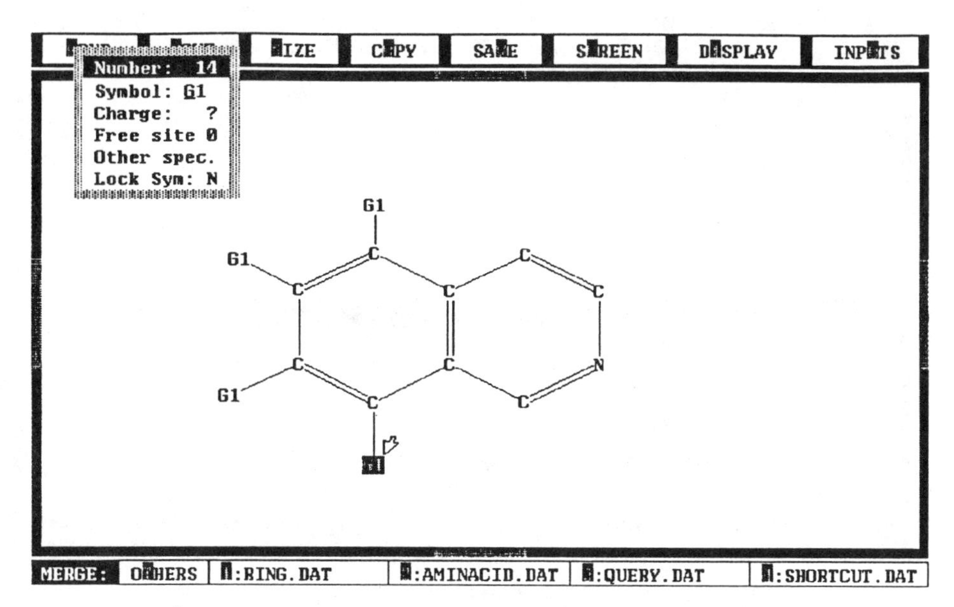

Figure 6. Adding G-groups to the structure.

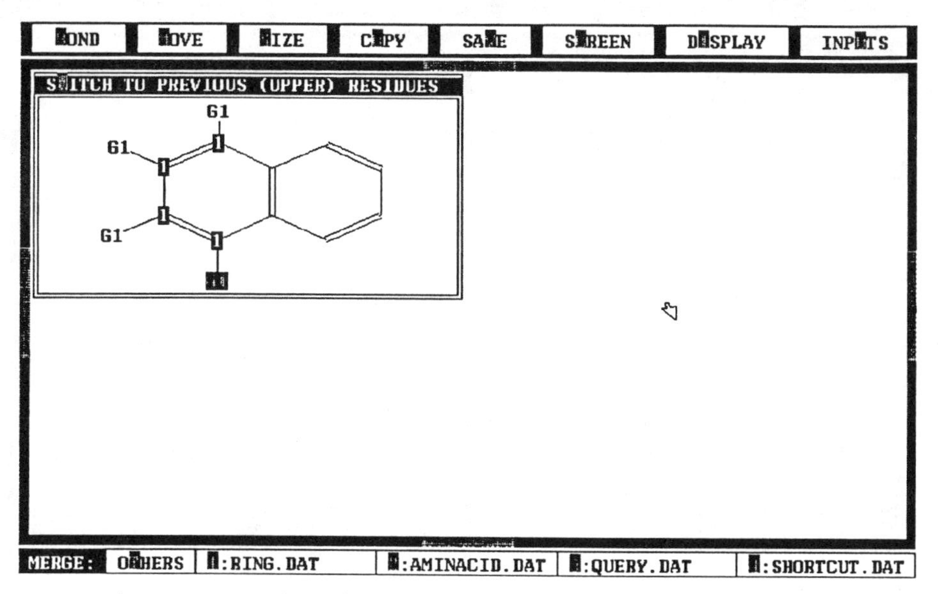

Figure 7. Edit window for defining G-group values.

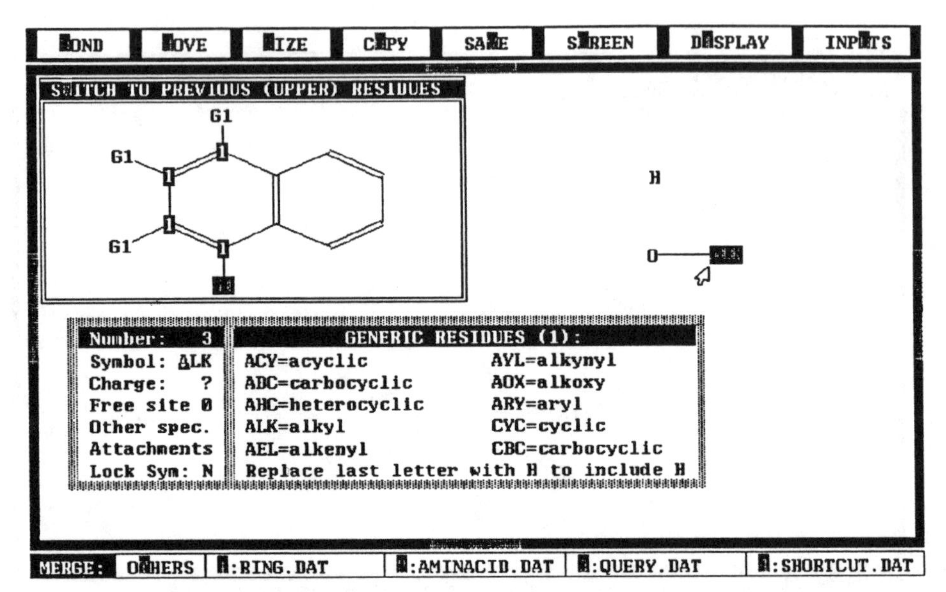

Figure 8. Generic residues help information.

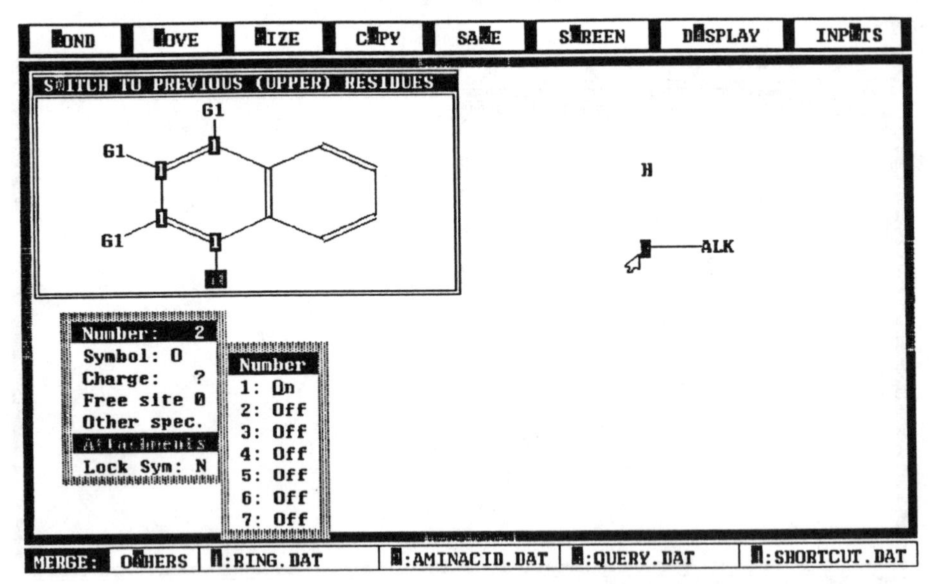

Figure 9. Specifying attachment points.

is checked and MOLKICK issues an error message if the string is incorrect, for example if the attachment points are not fully specified; this indicates that the query structure is not yet complete and cannot be searched by the S4 search system. When complete, the ROSDAL string is stored in the default file SUB.UPL.

ROSDAL is a form of description which has been designed specifically for communication of chemical structures between MOLKICK and S4 (10). Since the ROSDAL description is an ASCII string, this has a number of important advantages. Transmission of an ASCII string through a busy telecommunications network is less liable to error than the transmission of a binary file describing the graphics image of the structure. Also you can modify a ROSDAL string on your PC using your text editor if you wish; for this purpose you will need an understanding of the ROSDAL syntax, which is described in the MOLKICK manual and in the DIALOG user documentation "ROSDAL Manual for Users of the Beilstein Database at DIALOG". The latter documentation also describes how to write and submit correct ROSDAL strings without using MOLKICK, although this is not recommended for any but the simplest types of query structure.

At this point you are ready to submit your query structure to DIALOG. Switch to your communications program by typing the MOLKICK "hot-key" combination. If you are not already logged on to DIALOG and connected to the Beilstein database, File 390, then you should do this now, as described in the appropriate DIALOG documentation. Type QS at the DIALOG command prompt '?' to enter the Query Structure system which interfaces to S4. DIALOG displays the following or similar message

```
QS/QC Query Structure Version 1.01
Enter ROSDAL connection table
Line No.1 (enter '.@' to end or 'ABORT' to quit)
```

You now have two choices to submit your query. You can press the MOLKICK "query-key" combination, default ALT-R, and MOLKICK places the ROSDAL string into the keyboard buffer of your PC, complete with carriage returns and line feeds where necessary. DIALOG reads the ROSDAL string as if you had typed it at the keyboard. Alternatively, you can send the file SUB.UPL using the ASCII upload facility of your communications program, eg. ALT-U in CROSSTALK and PCPLOT. DIALOG reads the characters transmitted by the communications program until the ROSDAL terminator symbols '.@' are encountered, at which point S4 is started automatically.

After the structure search is complete, DIALOG displays the answer set number, the number of answers in the set and the QS command parameters which you specified

```
eg. S1 38 QS ID TEST
```

DIALOG now leaves the QS system and displays again the command prompt '?'.

You can also use MOLKICK to display on your PC the structures in the answer set, and print these locally on your printer if you wish. DIALOG sends structures to your PC as ROSDAL strings, again making use of the efficiency and security of ASCII file transmission. Since these are structures from the Beilstein structure file, each ROSDAL string also contains the coordinates of each atom, which enables MOLKICK to accurately display each structure in the form in which it was input to the Beilstein database. In the case of a stereoisomer, bond orientations are displayed which show the configuration of the molecule.

Use the DIALOG TYPE (T) or DISPLAY (D) commands with the appropriate display format in order to display the answers on your PC. Use the new GS or GR format with the TYPE command to view also the structure of each answer with or without stereochemistry respectively

```
eg. D S1/5/1-2
    T S1/5,GS/1-38
```

DIALOG sends the answers to your PC, where MOLKICK captures automatically each ROSDAL string in a temporary buffer within MOLKICK's memory. After all answers have been transmitted switch to MOLKICK by typing the "hot-key" combination, and the message

```
Display received hits (Y/N) ?
```

will appear in the centre of the MOLKICK screen, Figure 10. Typing Y causes the MOLKICK hit window to appear and the structure of the first answer is displayed, Figure 11. You can page through the answers using the "PREVIOUS HIT" and "NEXT HIT" options in the hit window, or alternatively use the PAGE-UP, PAGE-DOWN or cursor keys, Figure 12. After displaying or printing the structures, return to the MOLKICK editor screen by choosing the "BACK TO EDITOR" option. MOLKICK prompts you

```
Are you sure (Y/N) ?
```

If you type Y then MOLKICK flushes the buffer in which the ROSDAL strings were stored temporarily and returns you to the editor screen. Remember you can switch back to your DIALOG session at any time by typing the MOLKICK "hot-key" combination.

Since our query structure is still on the screen in the MOLKICK edit window, we can save the query into the file QUERY.DAT for future use, Figure 13. You may wish of course to combine this answer set with other search criteria, for example to select only those records where the alkoxyisoquinolines are used as the reactant in a preparation of some other isoquinoline derivative.

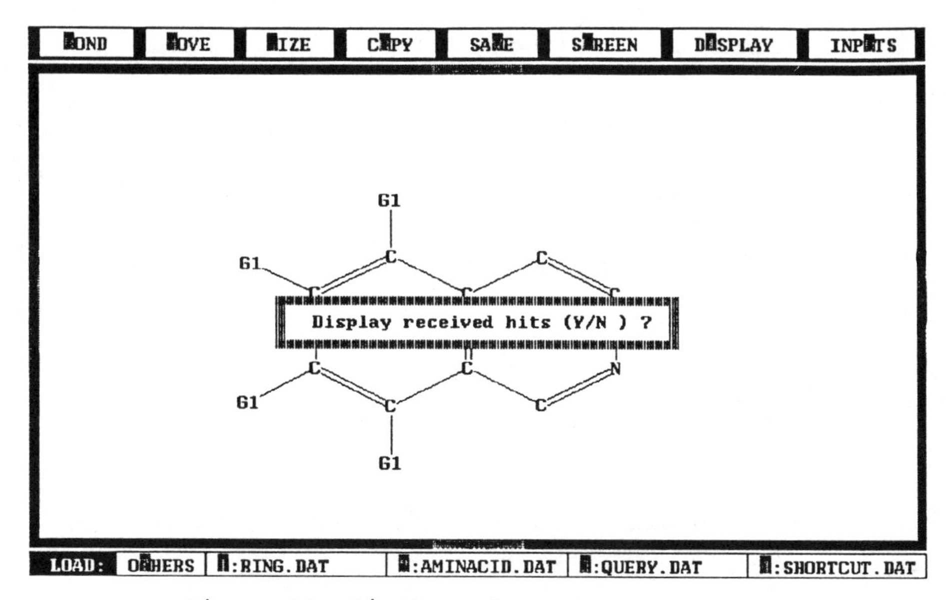

Figure 10. Display of answer structures.

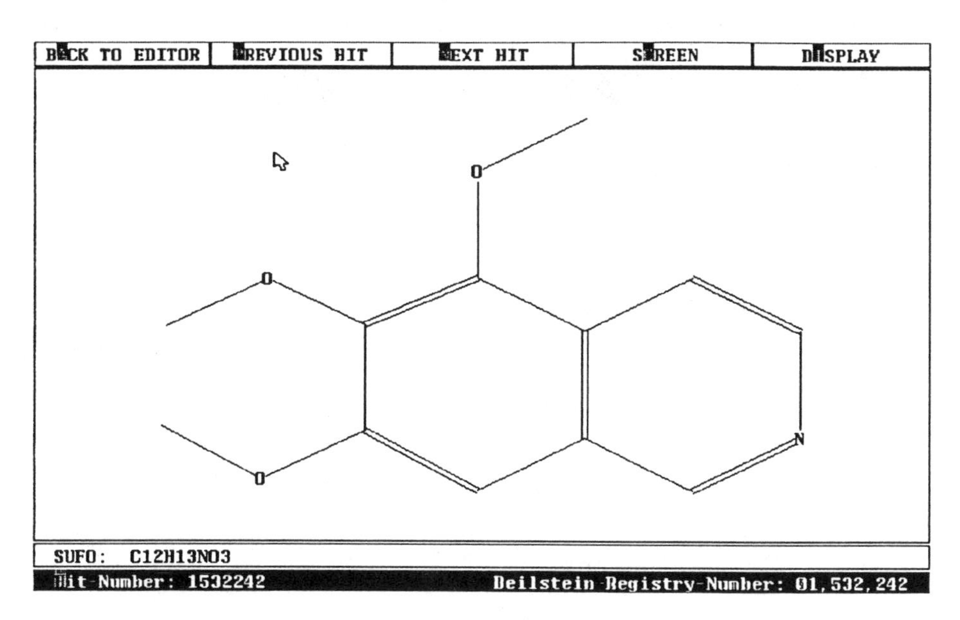

Figure 11. Display of first answer structure.

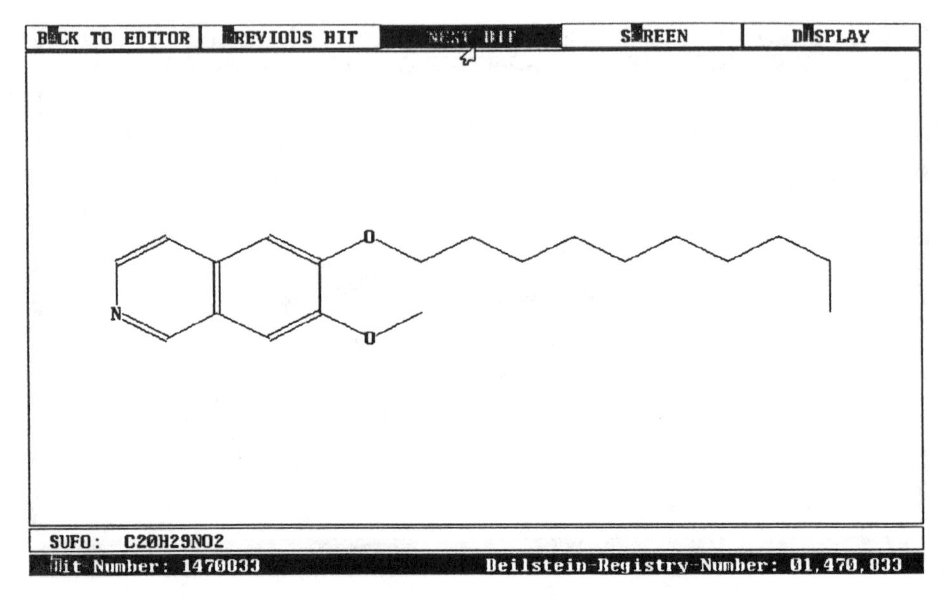

Figure 12. Display of subsequent answer structure.

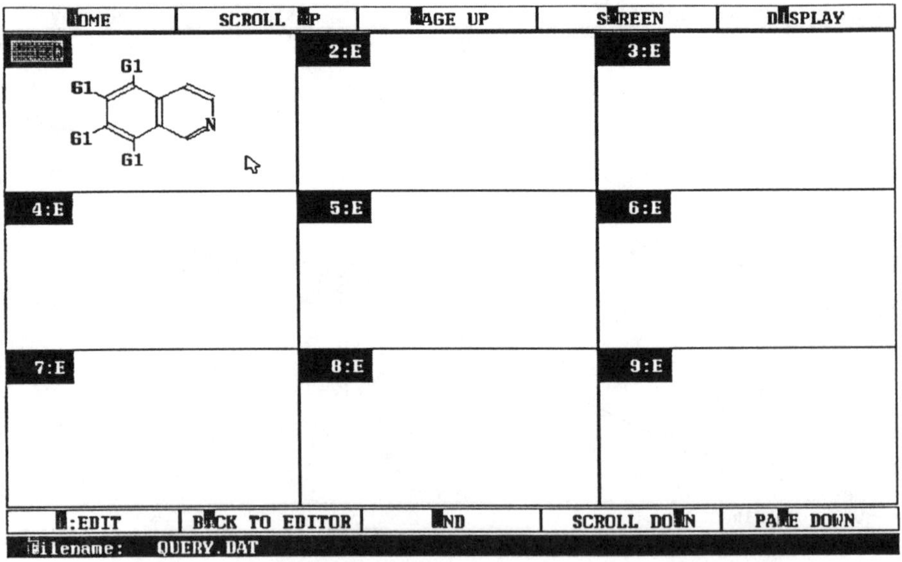

Figure 13. Saving query structure in the query file.

Summary

This chapter has presented a brief overview of structure searching of the Beilstein database on DIALOG using the S4 search system and the PC graphics-based query editor program MOLKICK. Online structure searching presents a new opportunity for DIALOG users, and is the result of significant development work by DIALOG, Softron GmbH and the Beilstein Institute. This development is continuing and will lead in time to the full implementation of search capabilities which are not yet realized, for example stereospecific and tautomer search. It has not been the purpose of this chapter to compare structure searching of Beilstein on DIALOG and on STN, since such a comparison would be premature and possibly misleading. If there is one message which you should take with you after reading this chapter, it is "Try it yourself and see!".

Acknowledgments

The author wishes to thank colleagues at Springer-Verlag, DIALOG, the Beilstein Institute and Softron GmbH for their assistance and comments in the preparation of this chapter.

Literature Cited

1. Jochum, C.; Background on the Beilstein Computer Activities; in this book.
2. Stobaugh, R. E.; J. Chem. Inf. Comput. Sci., 1989, 29.
3. Mockus, J.; Stobaugh, R. E.; J. Chem. Inf. Comput. Sci. 1980, 20, 18-22.
4. Barnard, J. M.; In Chemical Structures: The International Language of Chemistry; Warr, W. A., Ed.; Springer-Verlag: Heidelberg, 1988; pp 113-126.
5. Farmer, N. A.; O'Hara, M. P.; Database 1980, 3, 10-25.
6. Barth, A.; The STN Implementation of the Beilstein Factual and Structure Database; in this book.
7. Nagy, M. Z.; Kozics, S.; Veszpremi, T.; Bruck, P.; In Chemical Structures; The International language of Chemistry; Warr, W. A., Ed.; Springer-Verlag: Heidelberg, 1988; pp 127-130.
8. Chemical Structure Information Systems; Warr, W. A., Ed.; ACS Symposium Series No. 400; American Chemical Society: Washington, DC, 1989.
9. Chemical Structure Software for Personal Computers; Meyer, D. E., Warr, W. A., Love, R. A., Eds.; ACS Professional Reference Book; American Chemical Society: Washington, DC, 1988.
10. Barnard, J. M.; Jochum, C. J.; Welford, S. M.; In Chemical Structure Information Systems; Warr, W. A., Ed.; ACS Symposium Series No. 400; American Chemical Society: Washington, DC, 1989, pp 76-81.

RECEIVED May 17, 1990

Chapter 6

Support of Industry Information Needs

Norman J. Santora

SmithKline Beecham Pharmaceuticals, 709 Swedeland Road, King of Prussia, PA 19406

The Beilstein Online Database provides pharmaceutical chemists with a powerful tool to support their varied needs. This paper gives examples on the use of the database in assisting the medicinal chemist with a problem in drug design; the isolation chemist with the determination of the structure of a natural product; the chemical information chemist with a Markush search; and the organic chemist with a reaction search. The latter two examples demonstrate novel applications of the Lawson Number field code.

The examples make use of the variety and range-searching capabilities of the quantitative field codes available to the user of the Beilstein Online Database.

Medicinal Chemistry: Drug Design

A medicinal chemist would like to do a pattern recognition study on antibiotic candidates related to penicillin G **1**:

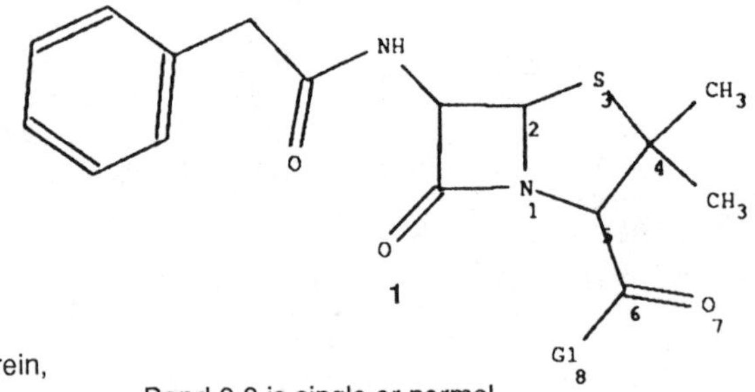

NH O S 3 2 4 CH$_3$ CH$_3$ N 1 O 5 6 O 7 G1 8

1

wherein,

Bond 6-8 is single or normal
Bond 6-7 is double or normal
All bonds are Ring or Chain
G1 = O or N, allowing for carboxylic acids, esters or amides

0097–6156/90/0436–0080$06.00/0

Pattern recognition, a branch of artificial intelligence, can be viewed as consisting of two techniques:

1. Preprocessing technique: Intended to reduce the number of molecular descriptors in the data matrix; then, to separate the compounds into classes based upon the reduced number of molecular descriptors in the data matrix.

2. Display technique: In which the results of the preprocessing technique are operated upon by a graphical program in order to discriminate between the classes revealed.

The compounds obtained from this search of the Beilstein Online Database will provide a supervised learning set in which some of the compounds will be tagged with known classifications with regard to antibiotic activity, and the primary objective is to develop a rule which classifies them correctly; then, to apply the same rule to classify the remaining compounds whose antibiotic activities are unknown.

Methodology. The method of choice for this pattern recognition study is that developed by Cammarata (1) to classify pressor agents. The compounds chosen are selected such that they will be superimposable upon the template **2** of atoms from penicillin G; wherein, the numbered atoms correspond to the atoms numbered in **1**. All of the compounds obtained from a substructure search of **2**, indicated as L1 in the Beilstein Online Database, will be superimposable upon **2**; thereby, enabling the structural features at those atoms to be expressed in terms of molecular descriptor codings. The latter operation results in a template-feature data matrix.

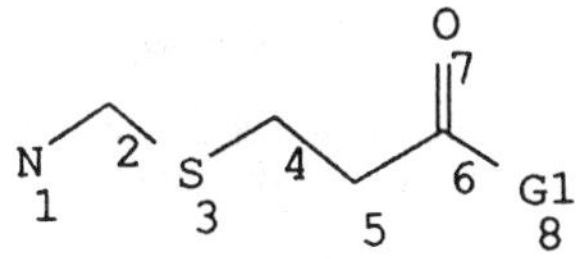

2

Cammarata has emphasized that molecular descriptor codings should be used, such as molar refraction, which can be employed as a measure of isosterism and bioisosterism.

The preprocessing technique chosen for this study is principle component analysis, in which the dimensionality of the template-feature data matrix is reduced. The principle component analysis determines the minimum number of molecular descriptor codings which will account for ≥ 80% of the variance in the data; thereby, allowing those molecular descriptor codings to represent the original template-feature data matrix.

Classification is accomplished by using a modification of the Andrews (2) technique for plotting data of more than two dimensions, in which the molecular descriptors for each molecule are transformed into variables which reflect the variance in the data as determined by the preprocessing step.

Search Of L1 In The Beilstein File

A full substructure search of L1 gave 3,273 answers:

L2 3273 S L1 SSS FULL

Restriction of L2 to records containing optical rotation information narrowed the answer set dramatically:

L3 180 S L2 AND (ORP OR ORD)/FA

wherein,

ORP is the Optical Rotatory Power field code
ORD is the Optical Rotation Dispersion field code
FA is Field Availability field code

The final answer set of 41 structures was obtained by requiring optical activity similar to that of Penicillin G:

L4 41 S L3 AND (175-310/ORP OR 300-550/ORD)

The diversity of structures obtained for the pattern recognition study **3**,**4** demonstrates the potential of this pattern recognition technique in enabling intuitive extrapolations to be made from the results of the classifications obtained.

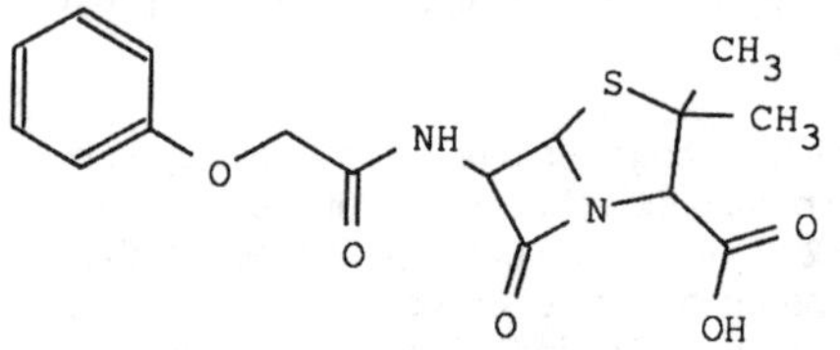

3 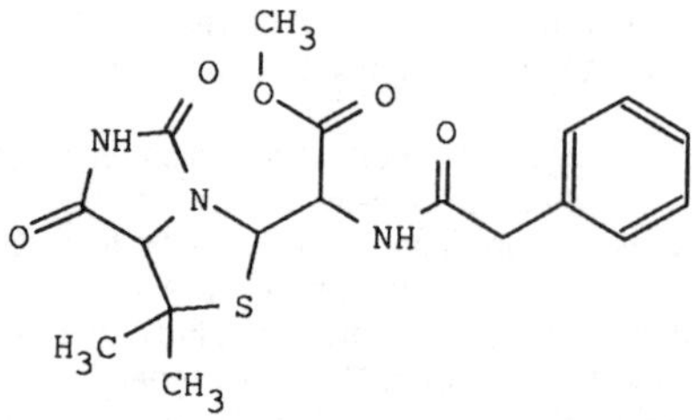 **4**

Isolation Chemistry: Identification Of A Natural Product

Many compounds isolated from one species are structurally related to compounds isolated from another species; for instance, compounds isolated from marine species are often similar in structure to compounds isolated from plants. Isolation chemists can make use of the Beilstein File to maximize the information gleaned about the structure of the compound in question from its initially-determined spectral and physical properties; in particular, because of the INP search field: Isolation from Natural Product.

A recent paper (3) described the use of single-crystal X-ray analysis to determine the structure **5** of an ethanol extract isolated from T. Purpurea, an Indian medicinal plant; however, the following example illustrates that the same information was available by making use of the earlier-determined physical and spectral properties of the compound:

MF = $C_{18}H_{14}O_4$
UVmax = 235 nm
Recognized as having a chelated OH group from red color with concentrated sulfuric acid.

Beilstein File Search. Although the use of search terms for UVmax and ^{13}C NMR were anticipated, the initial search statement resulted in a single hit:

L1 1 S $C_{18}H_{14}O_4$/MF AND INP/FA

wherein,

MF is the Molecular Formula field code
INP is the Isolated from Natural Product field code

A display of L1 in the STR HIT format revealed a diketone **6** isolated (4) from Pongamia glabra in 1942:

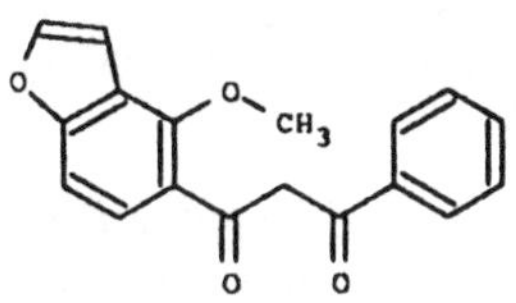

5 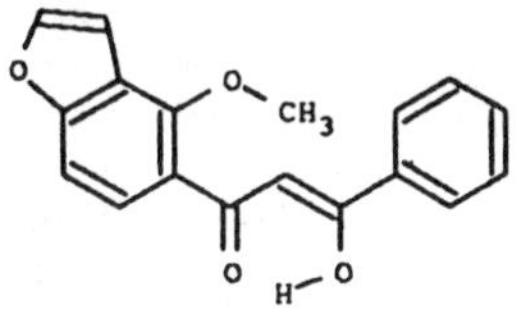 **6**

It is apparent that the latter authors may have isolated the enol form of the compound, **5**, since it was stated that the compound gave a ferric chloride color indicative of an enolic or phenolic hydroxyl group; however, that possibility was discounted since the compound doesn't undergo methylation or benzoylation reactions expected of such groups. Exact searches, in the Registry File, of compounds **5** and **6**, showed that Chemical Abstracts Services considers both compounds to be unique.

Chemical Information Specialist: Markush Search

This example, as well as the next example, utilize the Lawson Number search term in place of structures. Chemical information specialists are often required to due exhaustive priority searches of Markush structures. This example illustrates the usefulness of the Lawson Number search term to represent the Markush structure.

A search was requested for compounds represented by the Markush structure **7**:

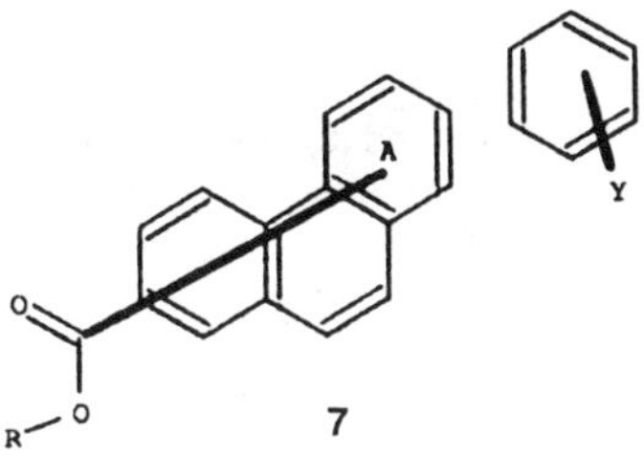

7

wherein,

A is a 6-membered ring bearing a N atom such that the fused ring system becomes a benzoquinoline.

R is C_{1-8} alkyl, straight-chained or branched, and the ester group can be attached to any of the rings of the benzoquinoline system.

Y is H, OH or OR'; wherein, R is C-alkyl, straight-chained or branched provided that the C-count of R + R' ≤ 8.

Rationale for Use of the Lawson Number. This Markush was considered too complex to be represented unambiguously in a strucural format for use in a substructure search; hence, it was decided to represent the Markush structure by a Lawson Number Range.

Robert Badger, of Springer Verlag, had revealed that the Lawson Number was equal approximately to eight times the System Number; thereby, making it attractive to explore the possibility of using the the latter relationship to provide a way of emulating the similarity search mode of Molecular Design Limited's REACCS database

Estimation of the Lawson Number Range of **7**. The Beilstein Institute's SANDRA program provided the System Number Range, SNrange, by inputting many different structural permutations and combinations of the variable groups in **7**. The highest value of the system number was 3346; whereas, the lowest value was 3268.

The Lawson Number Range, LNrange, equation was arrived at quite arbitrarily, after experimenting with correlations of system numbers of structures whose Lawson numbers were known. The study indicated that a broad similarity search would be possible, with confidence that the dictionary terms contemplated could reduce the initial large answer set to a reasonably sized final answer set.

$$\begin{aligned} \text{LNrange} &= (8.0 \pm 0.3) \times \text{SNrange} \\ &= (7.7 \times 3268) - (8.3 \times 3346) \\ &= 25164 - 27772 \end{aligned}$$

Selection of Dictionary Terms. Name segments and atom counts were chosen which would serve to restrict the anticipated large answer set resulting from the Lawson Number Range search.

A. Name Segments

1. BENZO
2. QUINOLINE
3. ESTER
4. PHEN?

wherein,

(1) and (2) provide for a variable benzoquinoline.
(4) provides for an ether linkage (phenyl) as well as OH (phenol)

B. Atom Counts

1. 1/N
2. 1 < O < 4
3. 20 < C < 30

wherein,

(3) is based on **7** (R = CH_3; Y = OH)

Beilstein File Search. A search of the Lawson Number Range provided an answer set, L1, larger than 63,000; however, the addition of the selected name segments, L2, and atom counts resulted in a final answer set,L3, of 16 structures.

L1	6333	125163 < LN < 27773
L2	25	L1 AND BENZO/CNS AND QUINOLINE/CNS AND ESTER/CNS AND PHEN?/CNS
L3	16	L2 AND 1/N AND 1 < O < 4 AND 20 < C < 30

Results. The structures obtained support the use of this technique for carrying out inexpensive, non-structural Markush searches; since, 50 % of the structures matched the requirements of the Markush perfectly; 38 % of the structures are similar; and 12 % of the structures are thought-provoking, at least. Examples of the structures which match the Markush **8**, are similar **9**, and provoke thought **10** are shown:

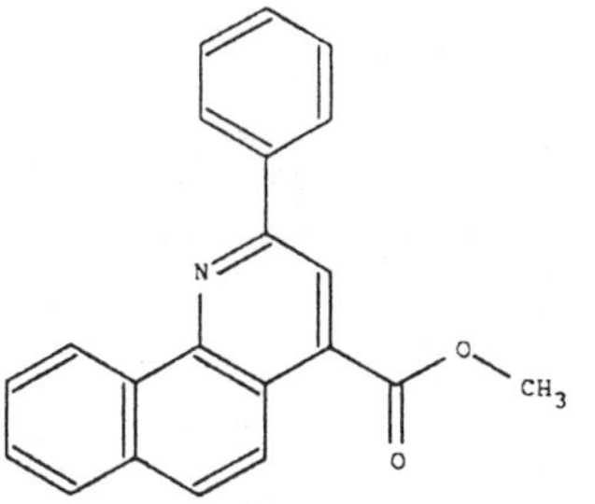

8 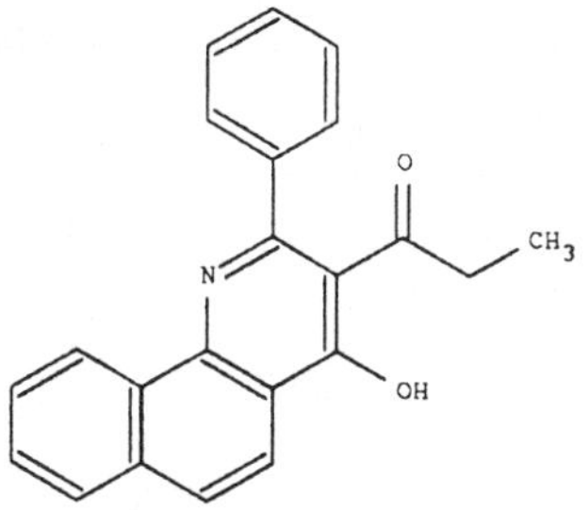 **9**

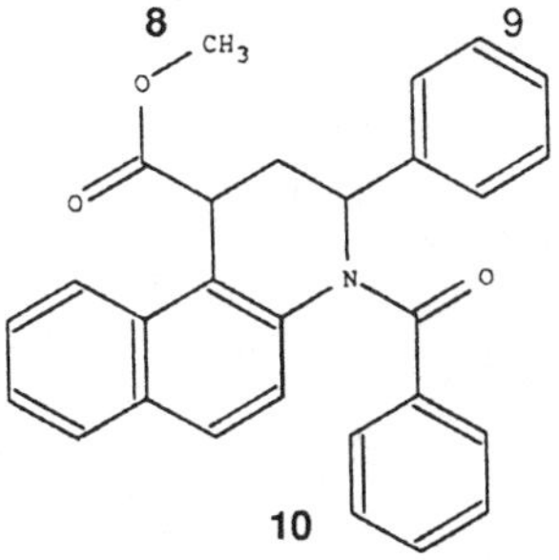

10

Organic Chemistry: Reaction Search

This example illustrates the use of the Lawson Number in place of the reaction product in the study of the Michael addition of aziridines to acrylamide, methyl acrylate or ethyl acrylate.

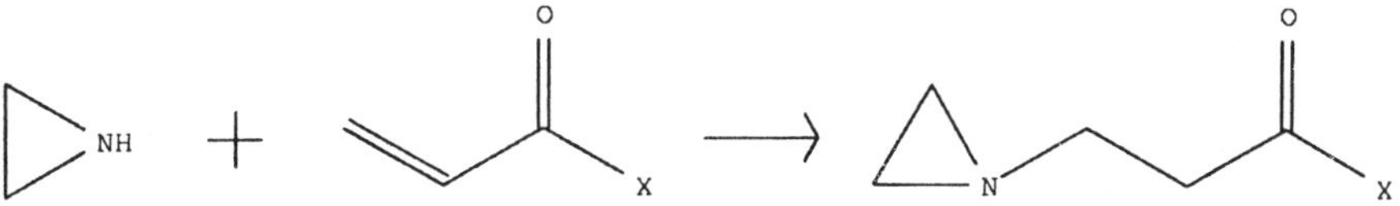

wherein,

$$X = NH_2, OCH_3, OCH_2CH_3$$

The use of a Lawson Number Range to represent a molecule provides the potential of performing similarity reaction searches. Similarity reaction searches can provide chemists with unexpected assistance in developing a scheme for carrying out an uncommon reaction.

Investigations with using the Lawson Number Range in place of the traditional terms: molecular formula, compound name, structure or Beilstein Registry Number, proved that the Lawson Number Range could not be used in the same manner as described in the Beilstein Database Manual; therefore, it was decided to use dictionary terms along with the Lawson Number Range.

Beilstein File Search. A search of the Lawson Number Range gave more than 17,000 answers, L1, which were reduced to the final answer set, L2, by the restrictions imposed by the dictionary terms for the name segments selected previously. The use of the (S) operator in L2 was to assure that the combined terms would be in the name of a single chemical substance:

```
L1    17631   S 23369 < LN < 25200

L2    10      S L1 AND (AZIRIDIN? (S) PROPION?)/CNS
```

Five of the ten structures, **11**, **12** and **13**, were relevant to the requirements of the search request:

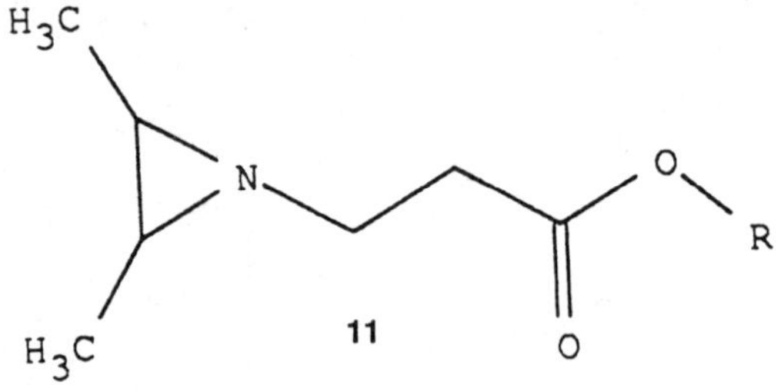

11

wherein,

R = CH_3 or CH_3CH_2

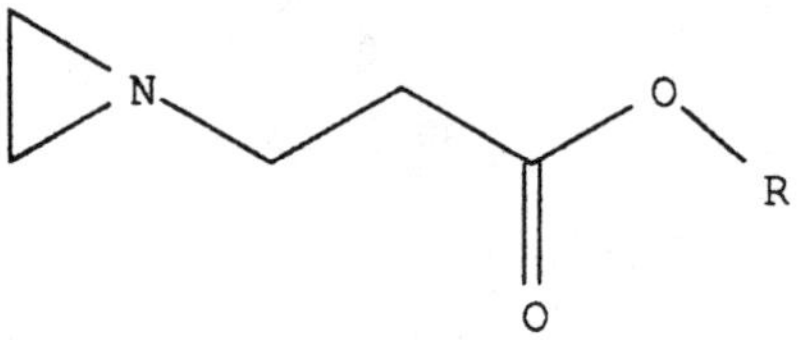

12

wherein,

R = CH_3 OR CH_3CH_2

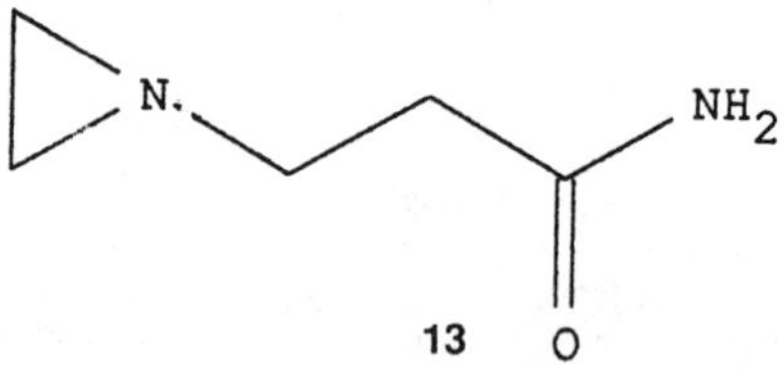

13

Conclusion

The Lawson Number Range was introduced as a viable representation of a similarity substructural unit, which emulates the REACCS similarity modes for substructure searching and reaction searching. It must be emphasized that the method of calculating the Lawson Number Range, introduced in this paper, is not to be regarded as being generally applicable; nevertheless, examples have illustrated their utility in performing Markush and reaction searches. Other examples have illustrated that the Beilstein Online File can aid medicinal chemists in drug design and isolation chemists in the identification of compounds isolated from natural products.

Literature Cited

1. Cammarata, A.; Menon, G. K. J. Med. Chem. 1976, 19, 739-748.
2. Andrews, D. F. Biometrics 1972, 28, 125-136.
3. Parmar, V. S.; Rathore, J. S.; Jain, R.; Henderson, D. A.; Malone, J. F. Phytochemistry 1989, 28, 591-593.
4. Rangaswami, S. Pr. Indian Acad. 1942, 15, 417-418.

RECEIVED May 17, 1990

Chapter 7

Searching for Chemical Reaction Information

Damon D. Ridley

Department of Organic Chemistry, University of Sydney, NSW 2006, Sydney, Australia

The Beilstein File provides excellent opportunities for the searching of information on chemical reactions from 1830 onwards. Information on preparations, or reactions, of a single substance or of a group of substances, using specific reagents, classes of reagents, or starting materials with common structural features can be obtained. Some searches are presented that illustrate simple solutions to very complex problems, while general techniques are suggested to help chemists solve their specific questions.

The Structure of the Full File

Figure 1 gives a diagrammatic representation of the information on a single substance in the Full File. This unique structure, in which reaction information is entered in fields within the record for the substance, permits searches for chemical reactions of a type not possible with other files.

Of the many search and display fields in the Beilstein File those with information on chemical reactions are listed in Figure 2. The INP (Isolation from Natural Product) and the CDER (Chemical DERivative) fields contain relevant information on the isolation of substances from natural products and information on simple chemical derivatives (for example oximes of carbonyl compounds, and acetates of alcohols) respectively. Generally these fields are displayed only when *all* available chemical reaction information is required; they are of lesser value as search fields for chemical reactions because of complications relating to choice of suitable search terms.

By far the most useful fields for searching chemical reaction information are the PRE (PREparation) and REA (REAction) fields. It is important to understand their structure and content from the outset, and to understand the entries differ in the Full File and the Short File.

The general procedure for finding reaction information in the File corresponds to the general method used in the Handbook. That is, it is necessary first to locate the records for the compounds of interest. In the

0097–6156/90/0436–0088$07.25/0

IDE

* A number of fields exist here for the identification of the substance.
* Most commonly substances are found through their names, molecular formulas, or structures

CHEM

* Within each record appear fields where chemical information can be searched and displayed.
* Most commonly the PRE and REA fields are used.
* Names (non–systematic) and name segments parsed only at spaces and punctuations of chemical reagents may be searched.
* Searches must be performed with code for field (and extension) e.g. furan/pre.sm

PHYS

* Within each record appear fields where physical data can be searched and displayed.
* Over 100 fields present – see database summary sheet.
* Most searches involve numeric searches, although general processes can be searched with controlled terms.

Figure 1. Schematic diagram of information in the Full File.

FIELD	DISPLAY CODE	ENTRIES FULL FILE	ENTRIES SHORT FILE
Preparation	PRE	430,000	2,200,000
Reaction	REA	63,000	310,000
Isoln. Natural Product	INP	4,700	30,000
Chemical Derivative	CDER	9,700	64,000

Figure 2. Some fields in the Beilstein File with chemical reaction information.

Handbook this is commonly done through searches on names or formulas. In the File, names of chemical substances appear in the CN (Chemical Name), SY (synonym), PRE and REA fields and search techniques must be *tailored for the field involved.* Groups of substances can also be obtained from searches relating to molecular formulas, Lawson numbers, the types and numbers of elements involved, molecular weights and through substructure searches.

The PRE Field

It is anticipated that eventually all entries will be in the form currently used in the Full File, and Figure 3 shows a typical entry which includes information in the PRE field (STN implementation). Entries in this field in the Short File are restricted to the *references only.*

(In this chapter are shown diagrams relating to the STN implementation. The Dialog implementation, discussed in chapter 4 , contains the same data provided by the Beilstein Institute although the display set out, display headings, and search commands are a little different. Implementations through other systems will have further differences. Nevertheless the differences can be correlated easily, and the concepts and techniques used specifically in this chapter are valid for the Database regardless of implementation).

Many of the substances originally entered in the Full File also appear in the Short File, so fields of interest for searching information on chemical reactions will contain entries with full data (for example names of substances involved in the reactions) and entries with references only.

In some searches for information on chemical reactions it may be necessary to restrict searches to entries in the Full File only and there are a number of ways in which this can be done. One way is to perform a RANge search based on Beilstein Registry Numbers. (RANge searches can be performed on all ONLINE networks. Procedures differ, and literature provided by the specific system should be consulted).

Another option is to include the search term "HANDBOOK" in search fields which allow the keyword suffix. Thus if L1 is an answer set of substances (for example from a substructure search) then the search:

=> S L1 AND HANDBOOK/PRE.KW

will retrieve Full File entries only.

In the typical entry in the PRE field (Full File) shown in Figure 3 it is to be noted that the field is further broken into subfields, characterized by subfield extensions. A correlation between the subfield codes and the way in which a typical preparation is visualised by the chemist is shown in Figure 4. Thus under the preparation of the registered substance are listed the starting materials in the PRE.SM field, and any 'inorganic' reactants, solvents or catalysts in the PRE.RGT field. (Note that either field codes PRE.SM (Starting Material) *or* PRE.EDT (EDucT) may be used from February, 1990. In the display of answers the subheading "Start" appears for this subfield). Names of substances in these fields can be searched in the *subfield* only, and either as the full word (as the 'bound phrase') or as a segment (segmentation of names in these fields in the STN implementation occurs at spaces and at punctuations only; other implementations have different rules).

BRN 1798193 Beilstein
MF C10 H14 Cl2 O4 Te
CN dichloro–bis–<2,4–dioxo–pentyl>–.lambda.4–tellane
Dichlor–bis–<2,4–dioxo–pentyl>–.lambda.4–tellan
FW 396.72
SO 2–01–00–00897
LN 1082

```
                 Cl
                 |
MeCOCH2COCH2TeCH2COCH2COMe
                 |
                 Cl
```

Preparation:
PRE
Start: acetylacetone, tellurium tetrachloride
Reag: chloroform
ByProd: telluracetylacetone dichloride
Reference(s):
1. Morgan,Drew, J.Chem.Soc. 121 <1922>,929, CODEN: JCSOA9
J.Chem.Soc., 125 <1924>,1602, CODEN: JCSOA9
Note(s):
2. Handbook Data

Figure 3. A sample record which includes display of preparation information.

THE PREPARATION OF COMPOUND [A]

Preparation Scheme *Search Using*

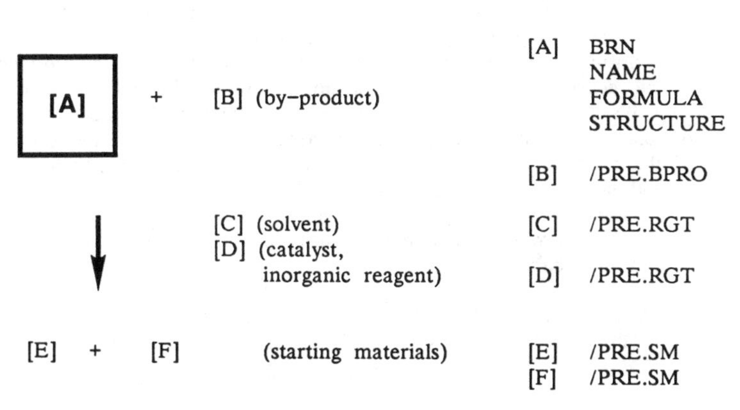

Figure 4. How preparations of compounds are indexed.

The entry shown in Figure 3 could be retrieved in a number of ways, depending upon the nature of the original question. More than likely the question may have required methods for preparations of dihalo–tellanes of the specific type illustrated. Thus the record may have been found through a substructure search and the preparation information displayed accordingly.

However this answer may have been found in other ways. For example the question may have been to find information on the reaction between acetylacetone and tellurium tetrachloride, or on the preparation of this dichloro–tellane specifically from acetylacetone. To answer either of these questions specifically it would be necessary for the search to include names of substances in the PRE subfields, whereupon the various rules for segmentations of names in these subfields need to be observed.

Thus the answer in Figure 3 would be retrieved with either:

```
=>  S ACETYLACETONE/PRE.SM AND TELLURIUM
    TETRACHLORIDE/PRE.SM
=>  S ACETYLACETONE/PRE.SM AND
    (TELLURIUM(W)TETRACHLORIDE)/PRE.SM
```

but not with searches involving any of the terms:

```
=>  S ACETYLACETONE/PRE
=>  S (ACETYL(W)ACETONE)/PRE.SM
=>  S TELLURIUM(W)TETRA(W)CHLORIDE/PRE.SM
```

There are subtleties here!

The REA Field

Again subfields with subfield extensions occur in the REA field, and Figures 5 and 6 show a typical record, and the correlation between the subfields and the way in which a typical reaction is visualised by the chemist, respectively. The rules for searching, and for segmentation, of names in the PRE subfields described above, apply to the entries under the REA field.

Again the nature of the question would in turn drive the search strategy, which in any case would have to be considered carefully. To illustrate a possible attack, suppose the question was "What happens when acetone cyanohydrin reacts with phosphorus pentachloride?"

So, consult Figure 6 and work out the roles of the various substances. The target record would be that for acetone cyanohydrin and this could be found by a search involving either the Beilstein Registry Number, the name, the formula, or the structure. Each of these has special considerations! The role of the phosphorus pentachloride would be as a "reaction partner", but what would be the appropriate search term?

To answer this the EXPAND command on STN (or the "thesaurus" feature on Dialog) would be used to identify possible search terms and the final search may be, *inter alia*:

```
=> S C4H7NO/MF AND (PCL5 OR PHOSPHORUS
     PENTACHLORIDE)/REA.RP
```

```
BRN  605391       Beilstein
MF   C4 H7 N O
CN   .alpha.-hydroxy-isobutyronitrile
     .alpha.-Hydroxy-isobutyronitril
FW   85.11
SO   3-03-00-00597; 0-03-00-00316; 4-03-00-00785; 2-03-00-00224
LN   1790
RN   75-86-5

      Me
      |
    MeCOH
      |
     NC

Chemical Reaction:
REA
     Part.:      PCl5
     Prod.:      .alpha.-chloro-isobutyronitrile,
                 .beta.-chloro-isobutyronitrile
     Reference(s):
     1.     Stevens, J.Amer.Chem.Soc. 70<1948>165,167, CODEN:
            JACSAT
     2.     Handbook Data
```

Figure 5. A sample record which includes display of reaction information.

THE REACTION OF COMPOUND [A]

Reaction Scheme		*Search Using*	
[A] +	[B] (reactants)	[A]	BRN NAME FORMULA STRUCTURE
		[B]	/REA.RP or /REA.RGT
↓	[C] (solvent)	[C]	/REA.RP
	[D] (catalyst)	[D]	/REA.RP
[E] + [F]	(products)	[E]	/REA.PRO
		[F]	/REA.PRO

Figure 6. How reactions of compounds are indexed.

Searches for Registered Substances – Full File

It is beyond the scope of this chapter to detail all the considerations relating to identification of substances, but clearly a variety of techniques may be used. Figure 7 gives some of the fields under those collectively called the substance identification fields and gives approximate numbers of entries within these fields. Entries in the CN field appear in English and in German (on separate lines) and relate to IUPAC nomenclature. However there are many difficulties with use of nomenclature, and in this case, as indeed in general, it is always advisable to make extensive use of the EXPAND (or EXPAND BACK) command to list search terms in any of the text fields. In this way appropriate search terms can be identified, and judicious use of truncation symbols can reduce search term costs.

Unless substances can be identified readily by use of a few search terms it is advisable to find substances by structure searches, either exact or substructure. Note that searches by structure are possible for the *registered substances*; searches by structure are not relevant for entries in the PRE or REA fields in which names, or part names are searchable.

Searches for Registered Substances – Short File

Also given in Figure 7 are the numbers of entries for substances which appear only in the Short File. Many have entries in the SY field, and all have entries in the MF and LN fields. All have structure data. While searches on molecular formulas can be of value in cases where few isomers are listed, they generally do not give answer sets of the precision required. Searches in the SY and LN fields can be limited also. In short, entries for substances can be retrieved reliably and comprehensively through searches involving structure alone.

Searches for Substances – in the PRE and REA Fields

Currently names of substances in the PRE and REA fields are not always systematic, so searches should be treated with caution. Again use of the EXPAND command to identify possible terms should be made extensively, and entries should be as general, and with searches involving as few terms linked together, as possible.

Some of these considerations have been mentioned previously particularly those relating to use of "synonyms" and segmentation, where special care needs to be exercised. Caution in linking together too many terms is mentioned not only because use of general terms helps overcome problems with nomenclature, but also because some questions are better asked more simply. Thus to search for phenyl Grignard reagents it makes more sense to use merely the term "phenylmagnesium" which in any case covers the chloride, bromide and iodide.

How to Find Preparations of a Single Substance

The options include:

* find the substance record and display the information in the

PRE field within this substance record
* look for the substance in the REA.PRO field
* look for the substance in the PRE.BPRO field.

The first of these is definitely preferred since
* data is evaluated and important preparations are in the PRE field,
* there are more entries in the PRE field than in the PRE.BPRO or REA.PRO fields,
* substances can be identified most reliably through structure searches and are more systematically named in the CN field than in fields under PRE and REA, and
* a further complication with searching for names in the PRE and REA fields occurs because of the parsing used. Thus a search:

=> S FURAN/REA.PRO

will retrieve answers for the preparation of furan itself, in addition to all those entries in which the segmented term "furan" appears in the name of the reaction product!

Of course much depends on the intent of the search and all possibilities should be attempted when maximum recall is required. It is to be noted also that:
* entries in the Short File give references only and do not list names of substances, and
* a display of the preparation field incurs a single display charge even though many preparations and reactions may be included.

Accordingly the most commonly used procedure to find the preparations of a single substance is to find the substance record and display the data in the PRE field.

How to Find Preparations of Related Substances

The only reliable method is to find the substance records through the appropriate substance identification field - commonly, but not exclusively, by substructure search. If required this answer set can be searched with the added term, PRE/FA, to limit retrievals to those which have information on preparation.

For example the search for all reports on the preparations of furocoumarins of the type (1) is outlined in Figure 8. 474 substances were found in the substructure search and of these 381 had data in the PRE field. It may be that all this data is not required, so the searcher is left with the problem of refining the search. The simplest way would be to define the substructure more precisely (for example if only hydroxyl substituents in the central ring are of interest, the suitable substructure could be built, searched, and the preparations displayed).

Alternatively, some key answers may be obtained by linking the answer set with starting materials of interest. In the case given in Figure 8 possible synthetic methods could start from the coumarin ring and then attach the furan, or from the benzofuran with subsequent attachment of the lactone.

FIELD	FULL FILE	SHORT FILE
BRN	ALL	ALL
CN	460,000	150
SY	1,000	1,400,000
MF	ALL	ALL
LN	ALL	ALL
STR	ALL	ALL
Names in PRE	430,000	NIL
Names in REA	63,000	NIL

Figure 7. Substance entries in the Beilstein File. (Approx. 3,000,000 substances – February 1990)

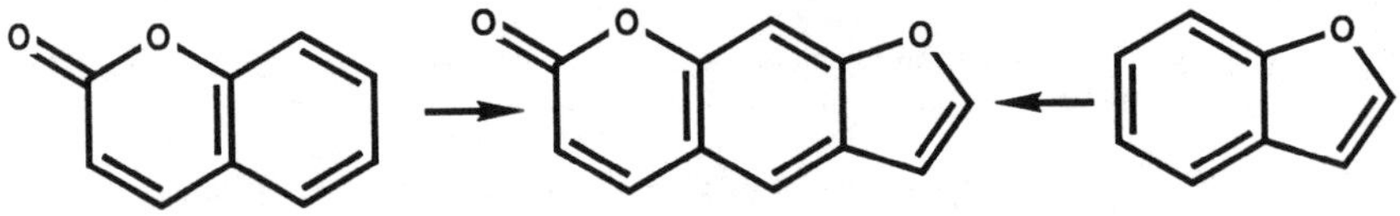

(1)

```
=>  S L1 SSS FUL
    L2    474   SEA SSS FUL L1

=>  S L2 AND PRE/FA
    L3    381   SEA L2 AND PRE/FA

=>  S L2 AND (COUMARIN OR CHROMEN?)/PRE.SM NOT
           FURO?/PRE.SM
    L4     22   SEA L2 AND (COUMARIN OR CHROMEN?

=>  S L2 AND BENZOFUR?/PRE.SM
    L5      6   SEA L2 AND BENZOFUR?/PRE.SM
```

Figure 8. Preparations of furocoumarins. L1 is the structure query based on structure (1).

The key to success in these types of searches is to identify (part) names of starting materials.

To answer the question relating to attachment of furan ring to the coumarin we would anticipate names of substances of the coumarin class in the PRE field to be listed as coumarin or chromenone (or chromen-2-one). Hence these search terms would seem appropriate. However many records would also include the name "furocoumarin" or "furochromenone" in the PRE.SM field and these would involve conversion of a furocoumarin to a related furocoumarin. Such answers would not be relevant to the search question, so the NOT operator should be used with the FURO?/PRE.SM term in the search.

Thus the search (L2 is the answer set of 474 substances from the substructure search):

=> S L2 AND (COUMARIN OR CHROMEN?)/PRE.SM NOT FURO?/PRE.SM

gave 22 answers which had a variety of nice methods for closure of the furan ring and Figure 9 gives an example.

The search:

=> S L2 AND BENZOFUR?/PRE.SM

gave 6 answers with 5 different methods of which one is shown in Figure 10.

These searches cannot be considered comprehensive, but they can give a quick entry into the literature and illustrate general techniques which can be used to retrieve valuable answers from the database.

How to Find Reactions of a Single Substance

Here the options are:

* find the substance record and display the information in the REA field within this substance record
* look for the substance in the PRE.SM field
* look for the substance in the REA.RP field.

The points raised in the preceding section are again relevant, although special caution may need to be exercised here on account of the greater number of entries under the PRE field than under the REA field. (It is worth noting that sometimes many more repetitions occur within the REA field and at times searches in the REA field may produce the greater number of relevant entries).

In order to assess which of the three options above may be preferable, consider the results shown in Figure 11. Thus when the record for the registered substance, furan, was recalled it was found that there were 247 entries in the REA field. That is, 247 different, and, as derived through the evaluation policies of the Beilstein Institute, significant reactions of the specific substance, furan.

However, because of the segmentation rules involving names of substances in this field, the search:

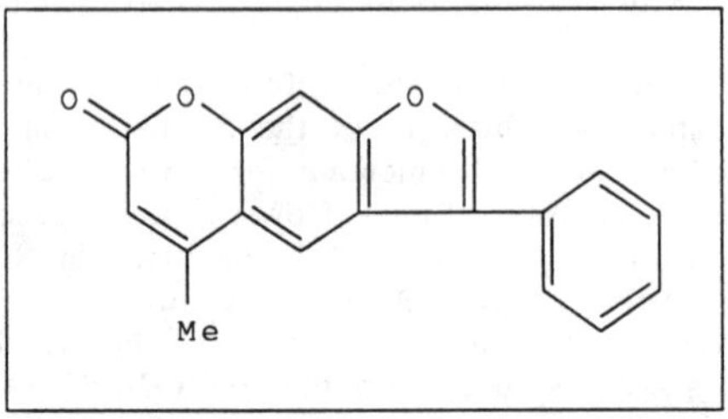

Preparation:
PRE

Start: 4-methyl-7-phenacyloxy-coumarin
Reag: sodium ethylate, ethanol
Detail: anschliessendes Behandeln mit wss. Salzsaeure
Reference(s):
1. Caporale, Antonello, Farmaco Ed.Sci. 13 <1958> 363, 366, CODEN: FRPSAX
Note(s):
2. Handbook Data

Figure 9. A sample answer for conversion of coumarins to furocoumarins.

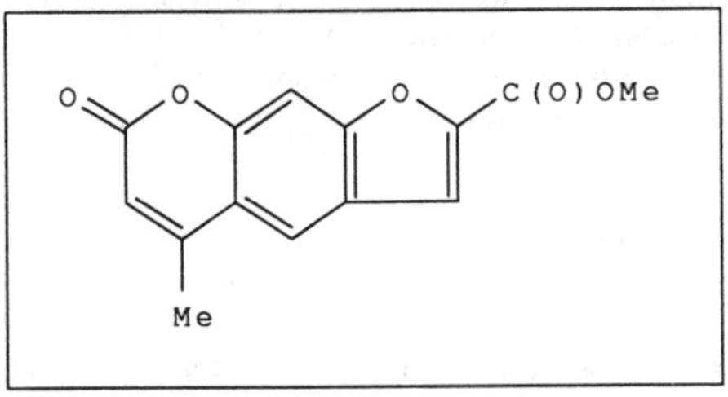

Preparation:
PRE

Start: 6-hydroxy-benzofuran-2-carboxylic acid methyl ester, 3-chloro-cis-crotonic acid
Reag: polyphosphoric acid
Reference(s):
1. Dann, Illing, Liebigs Ann.Chem. 605 <1957> 146, 152, CODEN: LACHDL
Note(s):
2. Handbook Data

PRE

Start: 6-<(.XI.)-2-methoxycarbonyl-1-methyl-vinyloxy>-benzofuran-2-carboxylic acid methyl ester
Reag: polyphosphoric acid
Reference(s):
1. Dann, Illing, Liebigs Ann.Chem. 605 <1957> 146, 152 CODEN: LACHDL
Note(s):
2. Handbook Data

Figure 10. A sample answer for conversion of benzofurans to furocoumarins.

=> S FURAN/PRE.SM

retrieved 6,100 records. Some of these would have related to furan itself, but many would have related to derivatives of furan.

Figure 11 shows search results for furan as a "reaction partner" (40 answers), and results of a variety of corresponding searches for pyridoxal, strychnine and two of the nitro-pyridines. These examples were chosen since they represent the broad cross-section of substances, from simple parent heterocycles to their derivatives to substances with "trivial" names.

From these results, and indeed from general experience in searching for information on reactions in the File, it would appear preferable to find the reaction information in the REA field under the substance record, but the nature of the substance, and the intent of the search may demand use of the alternatives. It is worth pointing out again here that the information is evaluated, so the most significant preparations and reactions are listed in the PRE and REA fields for the substance record. However, the merit of data evaluation always depends upon whether the evaluation matches your own!

Two other aspects are worth remembering for searches involving terms from the various subfields within the PRE and REA fields. The first is that the nomenclature is not systematic and this can best be accommodated by use of very general terms and of the (S) operator (which restricts proximity to a single substance name). The second is that it can be difficult to find individual substances in cases other than when the name is very simple. Thus names in these fields are searched either as the complete name or as fragments and both options may be considered. Some relevant entries are also included in Figure 11 for nitro-pyridines, searched as complete names in subfields within the PRE and REA fields. In these cases the search terms were found easily by use of the EXPAND command but problems arise with even slightly more complicated compounds. Interestingly, the search:

=> S (2(W)NITRO(W)PYRIDINE)/PRE.SM

gave 38 answers which of course included 2-nitropyridines with further substituents. Naturally it would depend upon the intent of the search as to whether these 38 answers were of relevance.

How to Find Reactions of a Group of Substances

Consider search strategies for a question seeking information on the conversions of 1,4-diketones to pyrroles illustrated in Figure 12.

Through PRE Field. From previous discussion it will be apparent that a substructure search of the pyrrole substructure followed by listing of entries in the PRE fields will be a preferred approach but in this case be prepared with rolls of computer paper, lots of time to sift through the data (much of it irrelevant), and for questions from your accounts' section about ONLINE costs!

Thus some hundreds of compounds with substructure (2) would be expected and many would have multiple entries for their preparations. Indeed when a substructure search on the pyrrole (Figure 13) was performed 1252 structures were retrieved and 1181 had entries for preparations.

Search Term	Entries (REA field)	Entries (/PRE.SM)	Entries (/REA.RP)
FURAN	247	6,100	40
PYRIDOXAL	123	102	40
STRYCHNINE	66	65	29
2-NITRO-PYRIDINE	9	2	0
3-NITRO-PYRIDINE	23	6	1

Figure 11. Searching for reactions. A comparison between fields.

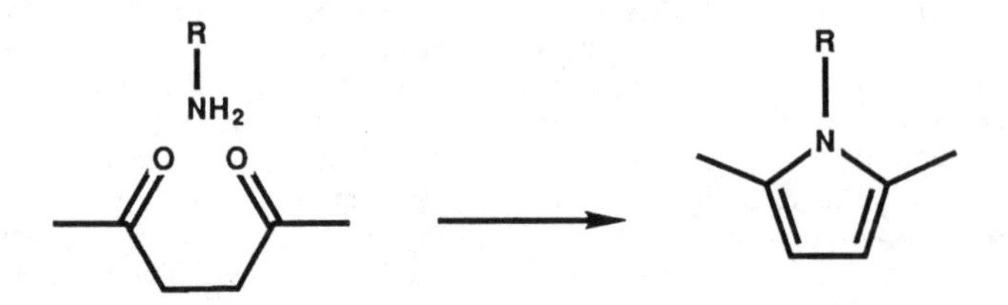

Figure 12. Conversion of diketones to pyrroles.

```
            E1
          2
 6 C   1   C     3  E1
     \   //  \
       C       C
       |       ||
     5 |       ||
       N ----- C
      /    4    \
 8 C              C 7
```

NODE ATTRIBUTES:

HCOUNT	IS E1	AT 2
HCOUNT	IS E1	AT 3
NSPEC	IS RC	AT 6
NSPEC	IS RC	AT 7
NSPEC	IS RC	AT 8

```
=>  S L6 SSS FUL
    L7    1252 SEA SSS FUL L6
=>  S L7 AND PRE/FA
    L8    1181 L7 AND PRE/FA
=>  S L7 AND HANDBOOK/PRE.KW
    L9     490 L7 AND HANDBOOK/PRE.KW
=>  S L9 AND (METHYLAMIN? OR ETHYLAMIN? OR
         CYCLOHEXYLAMIN? OR ANILIN?)/PRE.SM
    3141 METHYLAMIN?/PRE.SM
    2386 ETHYLAMIN?/PRE.SM
     287 CYCLOHEXYLAMIN?/PRE.SM
    9320 ANILIN?/PRE.SM
    L10  112 L9 AND (METHYLAMIN? ........)
```

Figure 13. Structure query (L6).

(At this stage those not thoroughly familiar with structure searching should not be too concerned with the details of the structure queries created in this chapter, and suffice it to say here that the query in Figure 13 had carbon atoms on the N- and at the 2- and 5- positions, with hydrogen atoms at the 3- and 4- positions. Nevertheless full knowledge of structure search techniques is essential for those involved in searches in chemistry, and in due course all users should take basic, then consider advanced structure search courses).

Of these answers, 691 had entries in the Short File only. The only option here is to display the PRE field for each and examine the original literature for answers of the type required.

Of the 490 entries in the Full File many will also have references for preparations alone (from the Short File entries) and again for comprehensive results the only option is to display the answers and refer to the original literature. However with these entries in the Full File effective searching for chemical reaction information can be performed since all will additionally have entries in the PRE.SM field.

Most answers of interest will have names of the starting diketones and of the amines, and it is only through use of the EXPAND command that suitable search terms can be identified. There are obvious problems with nomenclature of the diketones, although since full names and name fragments (segmented at punctuations remember) are indexed, some key diketones may be identified.

A more effective procedure would be to find a list of relevant amines by use of the EXPAND command. The modified search on L9 (the 490 structures in the Full File) shown in Figure 13 gave 112 answers and all were relevant. An example is shown in Figure 14.

Through REA Field. The alternative approach to the problem in Figure 12 would be to perform a substructure search on the starting diketone and to link this answer set with keywords in the REA.PRO field. A suitable search term may be PYRR?/REA.PRO. Care must be exercised here since the starting diketones may be entered under the structure of the enolic form, so the standard procedure, to search for the structure of the keto form and of the enol form, must be followed. Thus in Figure 15 are given the appropriate structure queries and the search, while a sample answer is given in Figure 16. It is interesting to note that this procedure produced 13 answers, and all were relevant.

As with all 'text' searches ONLINE, it is necessary to ascertain from the outset whether the most comprehensive set of answers, or a smaller key set of perhaps the most relevant answers is desired. Time and costs may be crucial considerations here. Until more precise entries, for example CAS Registry Numbers or Beilstein Registry Numbers are included in the PRE.SM field, it is realistic to aim merely for a number of key references and the search described is typical of the results that can be obtained.

Other considerations are to search for entries in the PRE.RGT field, to use truncation symbols to cover a number of search terms (without incurring multiple search term costs), and to use molecular formulas for simple 'inorganic' reagents.

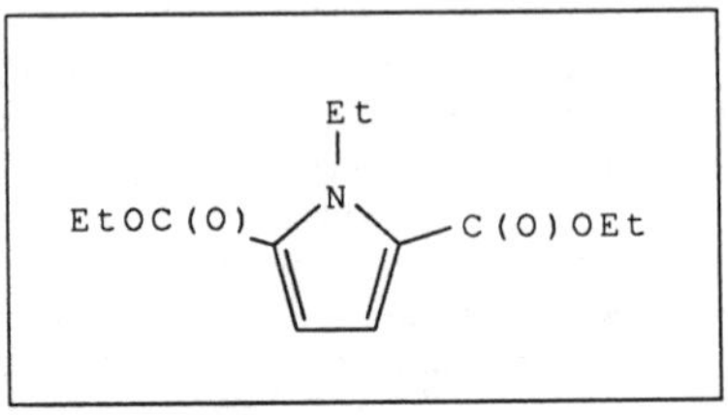

```
         Et
         |
EtOC(O)  N   C(O)OEt
```

Preparation:
PRE

Start: 2,5-dihydroxy-hexa-2,4-dienedioic acid diethyl ester, ethylamine hydrochloride
Reag: acetic acid, sodium acetate
Reference(s):
1. Kuhn, Dury, Liebigs Ann.Chem. 571 <1951> 44, 62, CODEN: LACHDL
Note(s):
2. Handbook Data

Figure 14. A sample answer for the conversion of diketones to pyrroles.

Diketone structure query (L11)

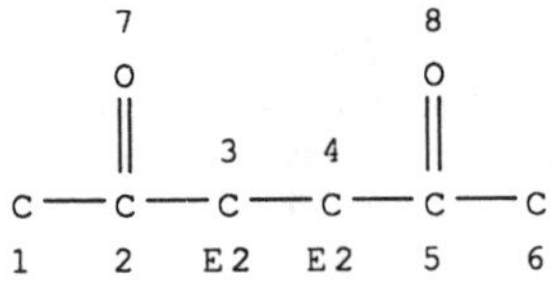

```
     7               8
     O               O
     ||   3    4     ||
C —— C —— C —— C —— C —— C
1    2    E2   E2   5    6
```

NODE ATTRIBUTES:

HCOUNT	IS E2	AT	3
HCOUNT	IS E2	AT	4
NSPEC	IS RC	AT	1
NSPEC	IS RC	AT	6

Dienol structure query (L12)

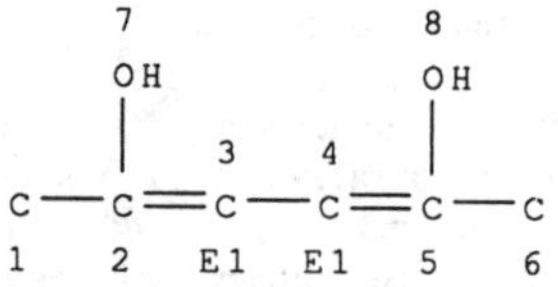

```
     7               8
     OH              OH
     |    3    4     |
C —— C == C —— C == C —— C
1    2    E1   E1   5    6
```

NODE ATTRIBUTES:

HCOUNT	IS E2	AT	3
HCOUNT	IS E2	AT	4
NSPEC	IS RC	AT	1
NSPEC	IS RC	AT	6

```
=> S L11 OR L12 SSS FUL
   L13        176 SEA SSS FUL L11 OR L12

=> S L13 AND PYRR?/REA.PRO
   L14        13 L13 AND PYRR?/REA.PRO
```

Figure 15. Conversion of diketones to pyrroles. Search queries and searches.

Careful Use of Structure Searches

While key results can be obtained for searching chemical reaction information by additions of search terms in the PRE.SM, PRE.RGT, and REA.RP fields, the problems with finding appropriate search terms must always be considered. At times it may be preferable to solve the question through structure searches alone.

Use of Combination of Fragments and Screens. Consider a request for the conversions of epoxides to episulfides (Figure 17). General options may include:

* substructure search on epoxide and (thiiran# or episulfide)/REA.PRO
* substructure search on episulfide and PRE/FA
* substructure search on episulfide and (epoxide or oxiran#)/PRE.SM
* search thiran/CNS and (epoxide or oxiran#)/PRE.SM
* search oxiran/CNS and (thiiran# or episulfide)/REA.PRO

Naturally the success of these substructure searches would depend on system limits, although RANge searches can always be used to obtain a complete answer set when a general structure fragment is required.

Alternatively, a key structural unit may be of interest and a substructure search may suffice. For example, consider a more specific search for preparations of all episulfides with the steroid nucleus.

The simplest approach would be to

* build structure of the steroid nucleus (3)
* build structure of the episulfide (4)
* search structure queries (3) AND (4)
* search these answers AND PRE/FA

This procedure retrieved 112 answers and preparations included compounds (5) and (6) (Figure 18). Interestingly there were no compounds with the thiiran attached in the chain at C–17 (e.g. part structure (7)). Those familiar with advanced structure search procedures involving the manual inclusion of 'screens' could, if desired, limit answers to only of the types (5), (6), or (7), or their combinations (CAS REGISTRY SCREEN numbers 1848, 1850, 1851).

Use of the VARiable command. As another example of the use of structure searches to solve questions on reactions, consider approaches to a request for information on the nitration of benzofuran (8). As with most searches, many

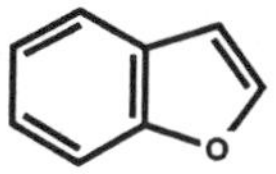

O

(8)

approaches are possible, but remembering that a preferred approach is through selection of substances by structure search then the combination of the answer set with entries in the PRE field, first thoughts would be directed towards a search for nitro benzofurans.

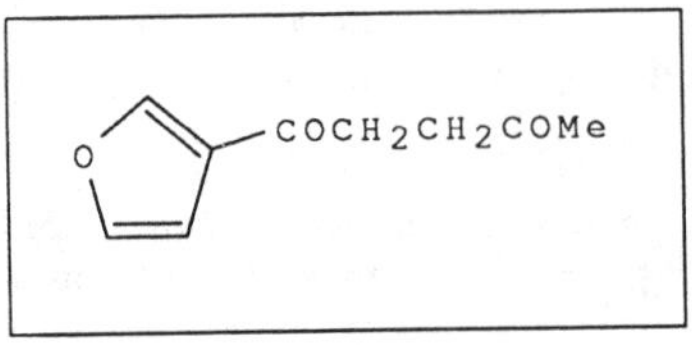

$COCH_2CH_2COMe$

Chemical Reaction:
REA

Part.: semicarbazide hydrochloride, potassium acetate, aqueous ethanol
Prod.: <2-<3>furyl-5-methyl-pyrrol-1-yl>-urea
Reference(s):
1. Kubota et al., Bull.Chem.Soc.Jpn. 38 <1965> 1191, 1192, CODEN: BCSJA8
Note(s):
2. Handbook Data

Figure 16. A sample answer for the conversions of diketones to pyrroles.

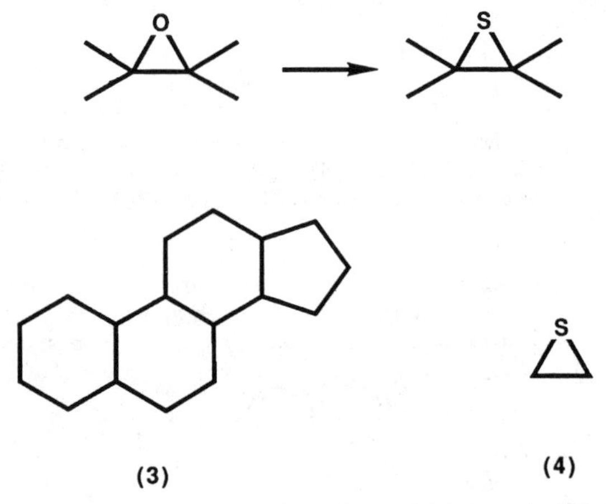

Figure 17. Conversion of epoxides to episulfides.

One option would be to build the benzofuran and the nitro group substructures separately and then perform a substructure search in which the two queries were ANDed together. This would produce an answer set where the nitro group was not necessarily attached to the benzofuran ring, but this would solve some nice problems, for example if selectivity in nitration was a prime consideration.

The easiest way to produce an answer set where the nitro group was attached to the benzofuran ring is to use the VARiable command and Figure 19 shows the structure query of the type needed. Thus you build the benzofuran, and a two node chain in which you specify the nodes as a variable (NOD 10 G1) and the nitro group (NOD 11 NO2). Then you ask that node 10 is one of the carbons at nodes in the benzofuran ring (VAR G1=1/2/5/6/7/8).

However both these answer sets will retrieve substances in whose records the actual nitration of the benzofuran ring is not included. For example some will be products of other reactions starting with nitrobenzofuran derivatives. To eliminate these it will be necessary to ensure that the starting material (in the PRE.SM field) is not a nitro compound so a search on the answer set (L16) from the substructure search:

=> S L16 NOT NITRO/PRE.SM

may be considered.

There still is one thing to remember - entries in the Short File do not have names of substances in the PRE field (only references) so they will remain in the new answer set. Thus it may be advisable to perform the substructure search only on those substances in the Full File, and for the heterocycles this means a RANge search up to Beilstein Registry Number 350,418. Alternatively, as shown in Figure 19, appropriate use may be made of the "HANDBOOK" keyword to limit the initial answer set to those with entries in the Full File.

It is to be emphasised that there would be other approaches, for example to include names of common nitrating agents (in PRE.SM,PRE.RGT) as search terms - but probably the intent of the search would be to find suitable nitrating agents!

Thus Figure 19 shows the search, and Figure 20 gives two of the answers. The first is an unexpected one! Other answers listed a number of other nitrating agents including aqueous nitric acid, nitric acid and acetic anhydride, mixtures of gaseous nitrogen oxides, reagents involving various nitrites and nitrates, and dinitrogen tetroxide.

Structure Searches. In summary, the point to be made here is that the searcher may best be served by defining the query very carefully, and by being aware that opportunites exist to make structure search queries very precise by full use of structure search capabilities. In short, time spent on learning how to search for substances by structure, can be repaid handsomely. It is worth pointing out here also, that, *inter alia,* the very elegant software package, STN EXPRESS, can assist greatly with the creation of the structure search query. Among the many structure building possibilities of version 3.0, for example, is one that permits variable points of attachment. The creation of the search question mentioned in Figure 19 can thus be done *very simply through an easily understood menu option.*

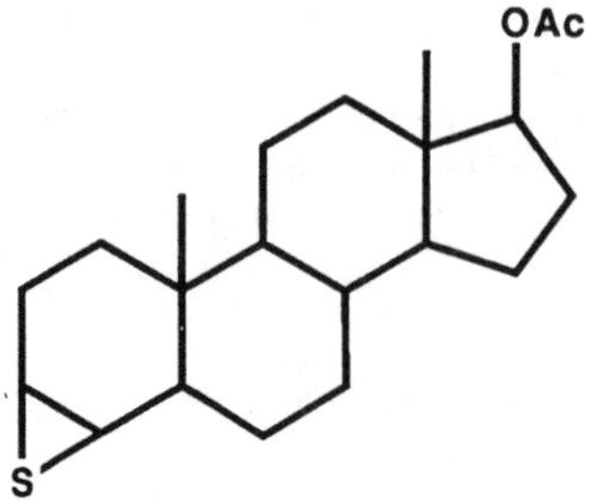

(5) 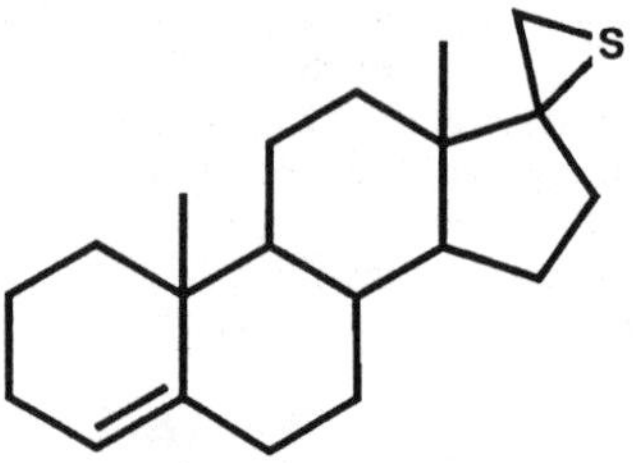 (6)

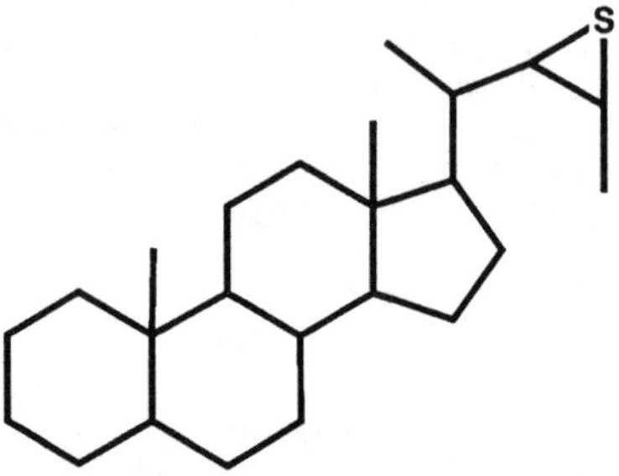

(7)

Figure 18. Steroid episulfides.

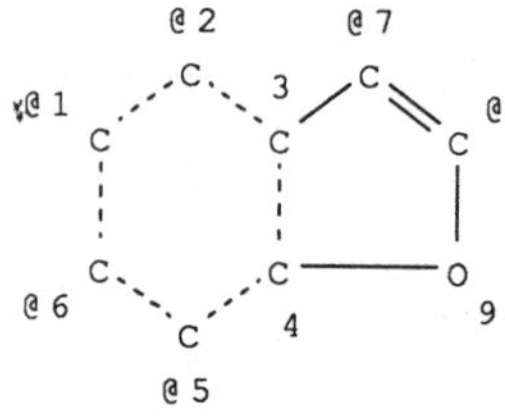

G1 - NO2
10 11

VAR G1=1/2/5/6/7/8
NODE ATTRIBUTES: NONE

```
=>  S L15 SSS FUL
    L16  316 SEA SSS FUL L15
=>  S L16 AND HANDBOOK/PRE.KW
    L17   51 L16 AND HANDBOOK/PRE.KW
=>  S L17 NOT NITRO/PRE.SM
    L18   20 L17 NOT NITRO/PRE.SM
```

Figure 19. Search query (L15) for nitration of benzofuran.

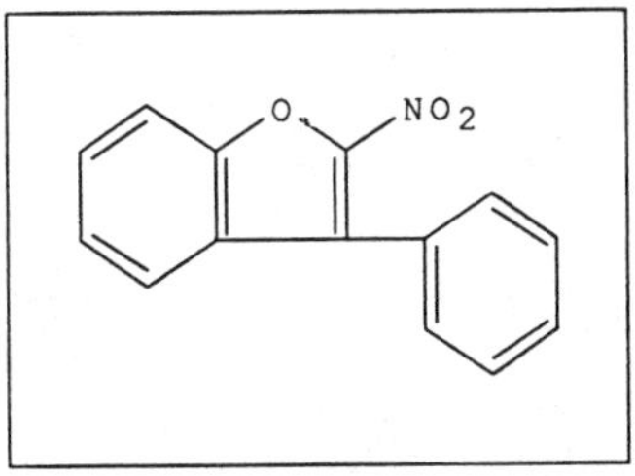

Preparation:
PRE

Start: 2-bromo-3-phenyl-coumarone, sodium nitrite
Reag: concentrated nitric acid, glacial acetic acid
Reference(s):
1. Stoermer, Chem.Ber. 44 <1911>, 1859 Anm. 2. CODEN: CHBEAM
Note(s):
2. Handbook Data

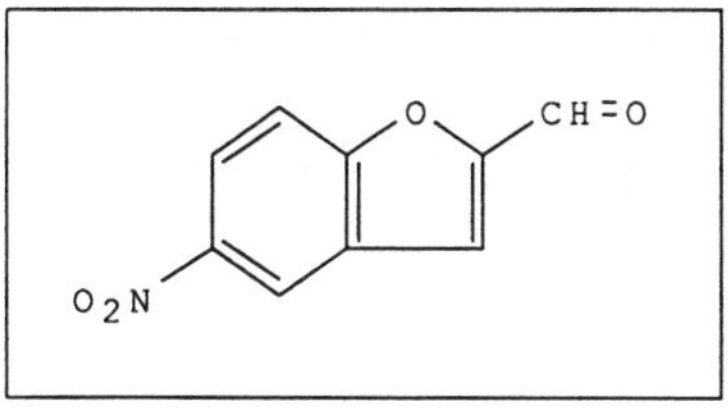

Preparation:
PRE

Start: benzofuran-2-carbaldehyde, nitric acid
Reag: sulfuric acid
Reference(s):
1. Kakimoto et al., Nippon Kagaku Zasshi 74 <1953> 636, Coden: NPKZAZ CA: 1954 12071
Note(s):
2. Handbook Data

Figure 20. Sample answers for nitration of benzofuran.

Chemical Reactions; *Chemical Abstracts* Files

The *Chemical Abstracts* Files. Figure 21 summarizes briefly some aspects of the Beilstein, CA, CAOLD and CASREACT Files. A most important consideration is the time period covered by the different files and the particular file of most value will sometimes be dependent upon this factor alone although it must be emphasised again that the nature and intent of the search are critical. Some of the "limitations" cited may well be advantages in individual types of searches!

Since chemical substances are indexed in the CA File as CAS Registry Numbers, the key to searching for chemical reaction information here is to identify the Registry Numbers of interest. In many instances this will be done in the Registry File and this answer set of substances can be crossed over to the CA File to find the bibliographic information.

If preparations of substances alone are required then use can be made of the fact that preparations are indexed under the Registry Number with suffix "P" added. Other searches can be performed through linking of the Registry Number with text terms, while some reactions can be identified by linking the Registry Numbers of starting material and required product. This last procedure may require two substructure searches in the Registry File, so costs may be significant.

An Example. To illustrate some of the features and limitations of the files, consider the question posed in Figure 22. Thus of interest is information relating to the preparations of caffeine analogues of type (10) in which the group "Q" is any atom other than carbon or hydrogen. In particular preparations that involve attachment of the imidazole ring to pyrimidines, for example (9), are required.

A search in the Beilstein File of the structure query shown in Figure 23 gave 1141 substances (answer set L20) of which only 264 appeared in the Full File. It is possible to search these with an appropriate keyword term and the search:

=> S L20 AND PYRIMID?/PRE.SM

gave 7 answers, one of which is shown in Figure 24. Interestingly only three different papers were listed but nevertheless this search produced very useful information quickly.

The remaining 877 substances appeared in the Short File and clearly there would be little value in displaying all the references, many of which would be repeated.

It is here where the complementary nature of the Beilstein and the CA Files could be used to advantage. Accordingly a search in the Registry File of the query in Figure 23 gave 2156 substances, the preparations of which are reported in 217 papers in the CA File. These papers would cover part of the period covered in the Short File of the Beilstein File and clearly in the first instance it would be preferable to display some of these CA File citations. Use of the CAOLD File may be considered also.

It may well be at this stage that the searcher has sufficient information on which to proceed, and further expense may not be justified. However to obtain more specific references to the question posed in Figure 22, it would be necessary to perform a substructure search on the pyrimidine (9). The

This table gives a brief summary only. Features/limitations refer to general situation, but will vary with nature and intent of search.

FILE	TIME PERIOD	NO.SUBSTANCES	SEARCH TERMS	FEATURES/LIMITATIONS
BEILSTEIN	1830–1959	500,000	Structure (preferred) names in /CN, /CNS /PRE.SM /PRE.RGT /REA.RP /REA.PRO	* time period covered * specific preparations/reactions listed under substance record; * data evaluated and validated; – may need multiple search terms for single substances; – references in answer set relate to different substances rather than different documents
	1960–1980	2,500,000	Structure (preferred)	
CA	1967–present	10,000,000	CAS Registry Numbers (preferred); can include text terms	* size of file * answer sets are documents * searches for classes of reactions possible – may need multiple search terms – general indexing is not reaction related
CAOLD	1960–1967		CAS Registry Numbers	
CASREACT	1985–present	900,000	CAS Registry Numbers	* single and multistep reactions * precise link between substances (reaction roles) possible – time period covered

Figure 21. A brief summary of some aspects of the Beilstein, CA, CAOLD, and CASREACT files.

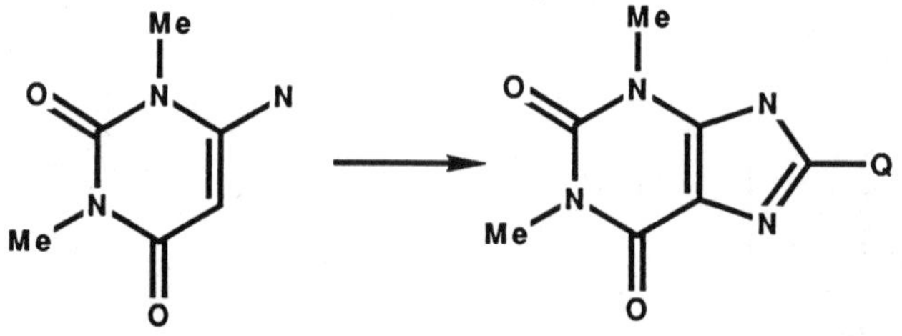

(9) (10)

Figure 22. Preparation of caffeine analogues from pyrimidines.

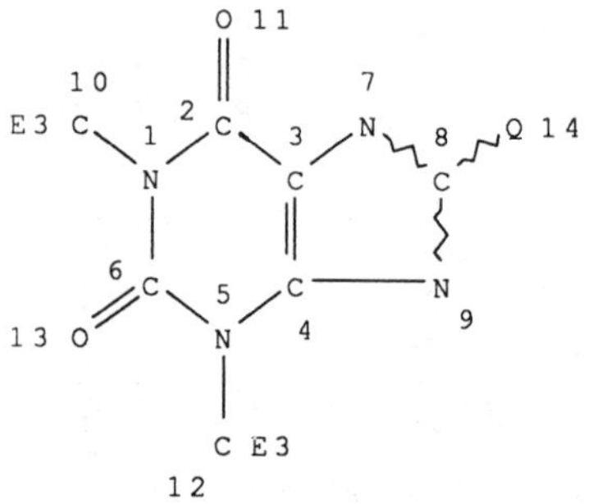

NODE ATTRIBUTES:

HCOUNT	IS E3	AT	10
HCOUNT	IS E3	AT	12
NSPEC	IS RC	AT	14

Figure 23. Preparation of caffeine analogues from pyrimidines. Search query (L19).

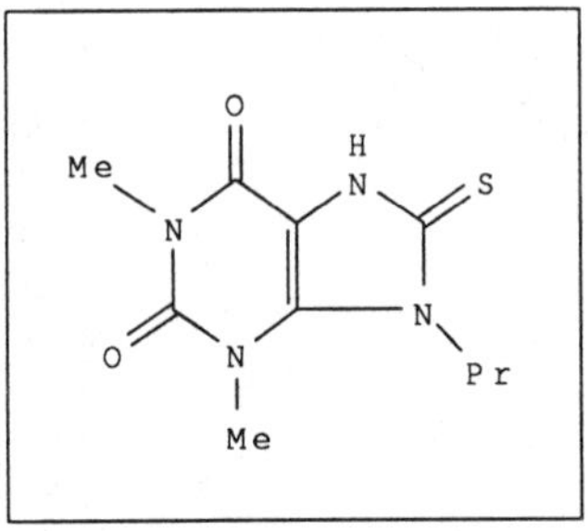

Preparation:
PRE

Start: 5,6-diamino-1,3-dimethyl-1H-pyrimidine-2,4-dione, propyl isothiocyanate

Reag: ethanol

Detail: Beim Erhitzen des Reaktionsprodukts mit konz. wss. HCl

Reference(s):

1. Blicke, Schaaf, J.Amer.Chem.Soc. 78<1956>5857, 5858, 5861, CODEN: JACSAT

Note(s):

2. Handbook Data

Figure 24. Preparation of caffeine analogues from pyrimidines. A sample answer from Beilstein File.

number of substances retrieved in the Registry File was 377, and when these Registry Numbers were crossed over to the CA File and searched together with the 217 papers mentioned above, 30 papers were found. All of these were relevant answers and Figure 25 shows the key index terms from a random two of them.

```
CA94(11):84066c  Synthesis and anthelmintic activity of some new
     1,3-disubstituted 1,2,3,6-tetrahydro-2,6-dioxo-8-purinyl carbamates.
     Shridhar, D. R.; Rao, K. Srinivasa; Bhopale, K. K.; Tripathi, H. N.;
     Sai, G. S. T. (Chem. Div., Indian Drugs Pharm. Ltd., Hyderabad 500
     037, India). Indian J. Chem., Sect. B, 19B(8), 699-701 (Eng) 1980.
     CODEN: IJSBDB. ISSN: 0376-4699.
KW   purinylcarbamate prepn anthelmintic
IT   Anthelmintics
        (substituted tetrahydrodioxopurinylcarbamates)
IT   1071-36-9
        (cyclization of, with (methoxyphenyl)isothiourea)
IT   5440-00-6   19677-97-5   50786-93-1   52998-22-8   76562-65-7
     76562-66-8
    (cyclization of, with bis(alkoxycarbonylmethyl)isothiourea,
    purinylcarbamate from)
```

```
CA84(21):150881m  Modified nucleoside syntheses. Ogura, Haruo;
     Takahashi, Hiroshi; Takeda, Kazuyoshi; Sakaguchi, Masakazu; Nimura,
     Noriyuki; Sakai, Hitomi (Sch. Pharm. Sci., Kitasato Univ., Tokyo,
     Japan). Hukusokan Kagaku Toronkai Koen Yoshishu, 8th, 154-8.
     Pharm. Inst., Tohoku Univ.: Sendai, Japan. (Japan) 1975. CODEN:
     32KOAD.
KW   nucleoside C; gluconyl isocyanate diamine condensation
IT   Amines, reactions
        (diamines, condensation with sugar isothiocyanates)
IT   Nucleosides
        (modified, from sugar isothiocyanates and diamines)
IT   1187-42-4   5440-00-6   6642-31-5   7318-00-5   13754-19-3
     29313-32-4   31295-41-7
        (reaction with sugar isothiocyanates)
IT   95-54-5, reactions
        (with sugar isothiocyanates)
```

Figure 25. Preparation of caffeine analogues from pyrimidines. Sample answers from CA File.

Conclusion

Over the years chemical scientists have developed many techniques for the searching for information on chemical reactions. Key entry points have been advanced texts, reaction reviews, monograph series, the **Beilstein Handbook** and *Chemical Abstracts*. The costs, ranging from purchases and then maintenance of libraries through to actual search time involved, may be very substantial and often underestimated.

With much information now available in computer readable form, the option exists for retrieval of information ONLINE. Again the costs may be substantial, although with knowledge of the database and good search techniques, the costs can be reduced. A feature of computerized searching, however, is that searches of a nature impossible through hardcopy searching can be performed readily.

The Beilstein File and the files produced by the Chemical Abstracts Service are complementary and, taken together, information on chemical reactions from 1830 to the present day can be retrieved. This chapter has illustrated some of the approaches, but is by no means exhaustive and chemists will find further ways to solve their specific problems.

Searching for chemical reaction information ONLINE is a challenge. With the certainty that the chemical literature will increase in size, it is a challenge which chemical scientists and their support library personnel must take up!

RECEIVED May 17, 1990

Chapter 8

Physical Property Data

Capabilities for Search and Retrieval

Andreas Barth

STN International, FIZ Karlsruhe, D-7514 Eggenstein–Leopoldshafen 2, Federal Republic of Germany

A comprehensive description of the different capabilities to search and retrieve physical properties in the Beilstein database is presented in this paper. The technics for searching numeric fields is discussed and the logic of numeric range retrieval is explained in detail. A new concept of range matching, called numeric range overlap detection, is introduced and illustrated with several examples. A new class of search fields have been developed which allow for more sophisticated property searches, e.g. like a restriction of the answer sets to critically evaluated (handbook) data. New Messenger capabilities supporting the special requirements of numeric databases are shortly described. Among them are the unit conversion feature, the capability to search for missing values and the concept of tolerance specification.

The Beilstein database is the largest source of physical property data with respect to both the number of substances and the number of different physical properties. There are about 70 properties indexed in numeric fields

0097–6156/90/0436–0113$06.00/0

and about 240 different keywords corresponding to physical entities stored in textual fields as controlled vocabulary. For all physical properties and controlled terms there is at least one literature reference pointing to the source of information. In Beilstein the physical properties are conceptually ordered in a hierarchical manner, i.e. like a thesaurus file. This ordering is also reflected in the order of the display formats as is depicted in Figure 1.

Physical properties are measured or calculated quantities recorded as numeric values associated with an uncertainty and a physical unit. The numeric values including the uncertainty are stored as numeric ranges in the database. There is an implicit physical unit corresponding to each search and display field which can be changed by the customer. In general, physical properties also depend on a set of parameters like temperature or pressure. With the Messenger software it is possible to perform parameter dependant searches of properties using a proximity operator.

The design of physical properties mimics the datastructure (for an example see table I). There is an entity name, e.g. Enthalpy of Formation, and a corresponding field qualifier HFOR. The field qualifiers for the parameters are built by using the abbreviation for the entity, a dot ('.') and an abbreviation for the parameter. Thus, the field qualifier for the Temperature (T) of the Heat of Formation is HFOR.T. The main qualifier, e.g. HFOR, is also the display format for this entity. To each numeric field a unique physical unit is associated, i.e. all values in a field are given in the same unit. In the following sections the retrieval of physical property information is described in detail.

Table I. Design of Physical Data

Field Name	Field Qualifier	Unit
Enthalpy of Formation	HFOR	J/mol
Temperature	HFOR.T	Cel
Pressure	HFOR.P	Torr

Format	Content
PHY	Physical Properties
CTUNCH	Controlled Terms of Unchecked Data
ECB	Electrochemical Behaviour
ELE	Electrical Data
MAG	Magnetic Data
MCS	Multi-Component System Data (MCS)
ADSM	Adsorption Data of MCS
ASSM	Association Data of MCS
BSPM	Boundary Surface Phenomena of MCS
ENEM	Energetic Data of MCS
GASM	Gas Phase System Data of MCS
LLSM	Liquid/Liquid System Data of MCS
LSSM	Liquid/Solid System Data of MCS
LVSM	Liquid/Vapour System Data of MCS
SOLM	Solution Behaviour of MCS
TRAM	Transport Phenomena of MCS
MEC	Mechanical Properties
OPT	Optical Data
SAG	State of Aggregation
CRY	Crystal Phase
GAS	Gas Phase
LIQ	Liquid State
SEP	Structure and Energy Parameters
CFM	Conformation
CPL	Coupling Phenomena
ELM	Electrical Moment
ELP	Electrical Polarizability
MEN	Molecular Energy
SKC	Skeletal Characteristics
SPE	Spectral Data
CTNQR	
EMS	Emission Spectrum
ESP	Electronic Spectrum
ESR	Electron Spin Resonance Spectrum
NMR	Nuclear Magnetic Resonance
OSM	Other Spectroscopic Methods
ROT	Rotational Spectrum
VIB	Vibrational Spectrum
THE	Thermodynamic Data
CAL	Calorific Data
HCP	Heat Capacity
THF	Thermodynamic Functions
TRA	Transport Phenomena
CND	Thermal Conductivity
DIF	Diffusion
VIS	Viscosity

Figure 1. Hierarchical structure of properties in Beilstein

Retrieval of Physical Property Information

Definition of Notations. In the Beilstein database, most of the physical properties are given as numeric data. Due to historical reasons, some properties are recorded as exact values without an uncertainty. These properties are called **single point entities** and they can be treated as simple numeric fields in bibliographic databases, i.e. only single values are stored. Physical properties which are recorded with an associated uncertainty are referred to as **numeric range entities**. A numeric range consists of a lower and an upper value which is equivalent to a value plus/minus an uncertainty. For these entities, the endpoints of the ranges are indexed in a numeric field.

In some cases, the knowledge about a property of a substance is very fuzzy and it is only known whether it is below or above a certain limit. As an example, some melting points are recorded to be greater than or equal to a specific value. In other words, the corresponding numeric range is an **open range**, since one endpoint is open (infinite). Ranges where both endpoints are finite are referred to as **closed ranges**. In general, it is of minor importance for a user whether the stored ranges are open or closed. However, there are some cases where it is necessary to include/exclude one type or the other and, therefore, it is important to understand the difference.

The concept of 'range searching' is associated with several different meanings. Firstly, the word 'range' could refer to both the query and the indexed range. In many databases it is possible to split the file into segments by ranges of field values. This feature is called **file segmentation**. A commonly used field for building these ranges in bibliographic databases is the Publication Year (PY). In this case, the user can specify a range of publication years using the SET command and all subsequent searches are limited to this range of years. This capability is often used to restrict the number of hits for queries with a huge number of intermediate answers. Without this segmentation, the search query would exceed the system limits and finally abort. In structure-oriented databases like the Beilstein or Registry file it is also possible to perform range searches using a range of Registry Numbers. These two possibilities are actually equivalent since they are both based on the primary file key, i.e. the Registry Number. In the case of the segmentation using the publication year, the range of years is converted into a range of registry numbers. The file segmentation feature will not be discussed further in this paper.

Another concept of 'range searching' is based on the formulation of a query as a numeric range. In numeric fields a search can be performed either as a search for a single value or as a search for a range of values. A simple example can be formulated for the field Publication Year (PY). For example, one could search for all documents published in 1988 (PY = 1988) or for those which have been published between 1986 and 1989 (1986 <= PY <= 1989). The latter example is called a search for a range of stored (indexed) values or a **numeric range search**. It is generally applied to search for a range of single values. Here, the notation 'range' implies a query range and not a retrieval of stored ranges. The numeric range searching capability is a general feature available to all STN databases with simple numeric fields containing single value entities like the Publication Year.

A more sophisticated feature has been developed to support the retrieval of physical properties associated with an uncertainty. Properties which are recorded as a value plus/minus an uncertainty are difficult to obtain in standard public information retrieval systems which are mainly focussing on the retrieval of bibliographic (text) data. The standard numeric range searching capability of Messenger has been enhanced to enable the user to retrieve the 'fuzzy' numeric data of physical properties. When the user performs a search of a 'fuzzy' physical property the Messenger software automatically invokes a procedure which performs an intersection between the query range and the stored (indexed) range. If the intersection is not zero the corresponding document is retrieved as a hit. This feature is called a **numeric range overlap detection** and it is actually based on an intersection between the query range (or point) and the stored ranges. It should be noted that the query could also be given as a single value, e.g. BP = 100. In this case, the software works in the same way as it does for a query range. The notation 'numeric range overlap detection' refers to the overlapping of the query range (or single value) and the stored range. Examples for the numeric range searching of physical properties are presented in the next subsections.

Strategies for Numeric Range Searching. All numeric physical properties can be searched using the standard numeric compare operators. In addition, the logical and proximity operators can be applied to numeric queries.

Table II. Set of operators applicable to numeric searches

Operator	Class	Use
AND, OR, NOT	Logical Operators	Combination of search fields
(L)	Link Proximity	Not applicable
(P)	Paragraph Proximity	Combination of parameters
(S)	Sentence Proximity	Not applicable
(W), (A)	Word, Adjacency Prox.	Not applicable
=, <= or =<, <, >= or =>, >	Compare operators	Numeric comparison of values

The set of allowed operators together with their application is shown in table II. All these operators are well known from the standard database retrieval, only the assignment of the (P)-proximity as a means to combine an entity with its parameters is unique to the numeric databases.

As an example to search for a numeric single point entity one may choose the Total Element Count (ELC). This field contains only single (exact) values and the standard numeric range search is performed. The strategy to search for a numeric range entity is the same as for a single point entity, however, the results of the search are different due to the overlap detection capability. In Figure 2 a schematic representation of a search for an exact value is shown. Here, a Boiling Point (BP) of 100 °C is searched. This is represented as a straight vertical line in the figure. The stored ranges of our example are represented as horizontal intervals. Those horizontal intervals which are intersected by the vertical line are comprising the answer set, i.e. the ranges 90-110, 95-115, and 100. There are two exact indexed values, 100 °C and 105 °C. Of course, only the first value belongs to the answer set.

In Figure 3 a similar picture is given for a numeric range search. The range (96 <= BP <= 108 °C) is represented by the shaded rectangle. As in Figure 2 we obtain the answer set as a result of the intersection between the query range and the stored ranges. The answer set for this example includes the ranges 90-110, 95-115, 98-100, 100, 105, 108-112, and >= 108. In this

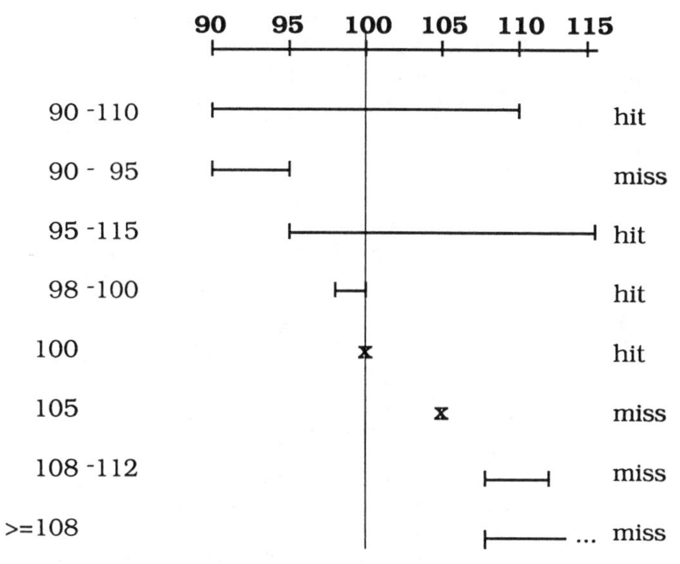

Figure 2. Schematic representation of a single value search

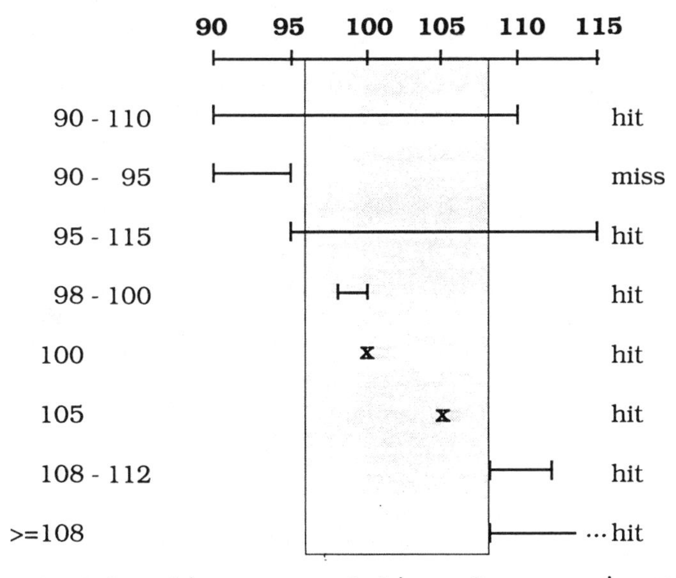

Figure 3. Schematic representation of a numeric range search

case the stored open range (>= 108) is also part of the answer set.

The last example (Figure 4) is a search of an open range, i.e. a Boiling Point BP >= 110 °C. Again, this interval is represented by a shaded rectangle but the right limit is at infinity (open). For this example the answer set contains the ranges 90-110, 95-115, 108-112, and >=108.

In the previous searches the answer set can become rather diffuse due the broadness of the stored ranges and it may be necessary that the answer set has to be reduced by combining it with other factual information. However, it is important to note that the answer set is complete in the sense that no possible hit will be missed. Even an imprecise specification like Melting Point MP >= 50 °C will be found by every query overlapping with the indexed range from 50 to infinity, e.g. a search for a Melting Point MP = 1000 °C will also retrieve this 'range'.

<u>More sophisticated property searches</u>. The standard numeric range searching with the overlap detection feature yields, in general, rather satisfying results. However, it is sometimes necessary to obtain more specific answer sets. For this purpose, a new subfield has been introduced for all physical properties containing a number of keywords describing the type of

Table III. List of Keywords for Physical Properties

Keyword	Meaning
HANDBOOK	critically evaluated data from the handbook
UNCHECKED	non-evaluated data from the literature excerpts
EXACT	single point value
RANGE	numeric range value
CLOSED	numeric range with finite boundaries
OPEN	numeric range with an infinite boundary
EXPERIMENTAL[1]	experimentally determined value

---

1 Currently all physical property values in the Beilstein database are experimentally determined.

data. This subfield is addressed by using the main field qualifier for the entity and concatenating it with '.KW' (Keyword), e.g. for the Boiling Point (BP) the abbreviation for this field is BP.KW. It is only searchable and may contain any subset of the keywords listed in table III.

It is obvious that these keywords are grouped in pairs, e.g. a numeric property may originate from the handbook or from the literature excerpts, it could be either an exact value or a range and the range could be either closed or open. The keyword 'EXPERIMENTAL' has been introduced for future use in case that the Beilstein Institute decides to include theoretically computed properties in their database. Using these keywords, it is possible to obtain more specific results than in the case of a pure numeric range search. The keywords have to be interpretated formally as a parameter of the physical property, hence, they must be combined with the (P)-proximity.

In Figure 5 an example is presented to restrict the numeric search to data from the Beilstein handbook using the keyword 'HANDBOOK'. The result from this search is obtained from critically evaluated data only. However, there may be other measurements for the substances of the answer set which are in the same range but originate from non-evaluated data. It may also be that there are completely different values for these substances which do not intersect at all with the numeric range of the query. The answer set is built by mapping the query and the stored numeric values and the keywords and the result is a set of Beilstein substances satisfying the query.

An interesting point is the restriction of searches to exact values. As described in the previous subsection a numeric search is performed by invoking the overlap detection procedure and the result comprises all the ranges intersecting the search query. Using the keyword 'EXACT' it is now possible to restrict the search to a single exact value. In figure 6 an example for such a search is presented. If this keyword is used in conjunction with a query range, the result is a numeric range where both endpoints are exactly identical to the endpoints of the query.

The last example in Figure 7 represents a possibility to exclude open ranges from the numeric range overlap detection. This means that the diffuse ranges are not taken into account when the answer set is built. In this case one has to search for exact values ('EXACT') or closed ranges ('CLOSED'). Analogously, it is also possible to restrict the search to retrieve only

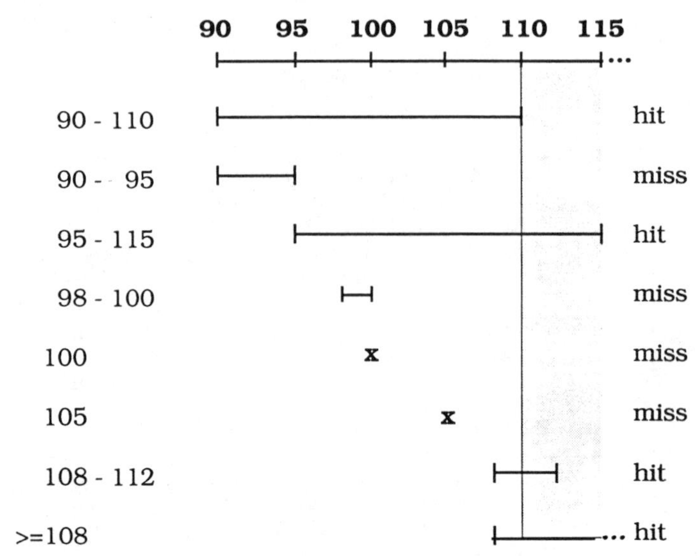

Figure 4. Schematic representation of an open ended range search

```
=> s 3 - 5/orp (p) handbook/orp.kw
          2256 3 - 5/ORP
         33253 HANDBOOK/ORP.KW
L5         892 3 - 5/ORP (P) HANDBOOK/ORP.KW

=> d hit

L5  ANSWER 1 OF 892

Optical Rotatory Power:
ORP  3.550 deg
     Type:   <alpha>
     Solv:   aq. NaOH
     Wavel:  589.00 nm
     Temp:   20.0 Cel
     Reference(s):
     1. E.Fischer, Chem.Ber. 40 <1907>,1758, CODEN: CHBEAM
     Note(s):
     2. Handbook Data
     3. 0.325 g in 4.0908 g Loesung.
     4. <d-.alpha.-bromo-isocaproyl>-hexaglycylglycine
```

Figure 5. Search example to restrict the answers to evaluated (Handbook) data

```
=> s 1650 - 1700/irm (p) exact/irm.kw
          4786 1650 - 1700/IRM
           190 EXACT/IRM.KW
L1          29 1650 - 1700/IRM (P) EXACT/IRM.KW

=> s 1650 - 1700/irs (p) exact/irs.kw
          9040 1650 - 1700/IRS
            80 EXACT/IRS.KW
L2           8 1650 - 1700/IRS (P) EXACT/IRS.KW

=> s l1 or l2
L3          37 L1 OR L2

=> d  hit

L3  ANSWER 1 of 37

Infrared Maximum:
IRM  1667 cm**-1
     Reference(s):
     1. Aelion, Champetier, Bull.Soc.Chim.Fr. 1949 529, CODEN: BSCFA
     Note(s):
     2. Handbook Data
     3. Absorption.
```

Figure 6. Search of exact values

```
=> s 20 - 30/mp (p) (exact or closed)/mp.kw
          4306 20 - 30/MP
        396045 EXACT/MP.KW
        819485 CLOSED/MP.KW
L4        4211 20 - 30/MP (P) (EXACT OR CLOSED)/MP.KW

=> d cn hit

L4  ANSWER 1 OF 4211

CN   di-undec-10-enoyl peroxide
     Di-undec-10-enoyl-peroxid

Melting Point:
Value (MP)               Solv.(MP.SOL)            Ref.  Note
(Cel)
-------------------------+------------------------+-----+-----
23.00 - 24.00            benzene, ethanol         2     1
23.00 - 24.00            benzene, petroleum ether 2     1

Reference(s):
2. Cooper, J.Chem.Soc. 1951 3106,3112, CODEN: JCSOA9

Note(s):
1. Handbook Data
```

Figure 7. Search to exclude open ranges

open ranges. It should be noted, however, that open ranges occur only for a very limited number of fields.

Finally, it should be noted that there is an additional subfield for physical properties containing the interval length. This is the difference between the upper and lower limit of the numeric range which is equal to the uncertainty of the measurement multiplied by two. The new field is built from the main qualifier concatenated with '.RAN', e.g. BP.RAN for the corresponding subfield of the Boiling Point. An EXPAND of this field shows the precision of the measurements for this property.

Specification of Tolerances. A search query for a numeric range can be specified either as a range or as a value plus/minus a tolerance. Both forms are treated by the Messenger software as equivalent, i.e. the following queries will result in the same answer set:

```
=> search  4 - 10 /ri
=> search  ri = 7 +- 3.
```

In the latter case, the search query is first transformed into a range and then the numeric range is searched. As described above, the answer set comprises all indexed ranges overlapping the range. The tolerance may be expressed in one of the following ways:

- as an absolute value, e.g. 132.09 +- 0.02 or
- as a percentage, e.g. 45 +- 2%.

It should be noted that the physical units for the value and the tolerance must be identical, a mixing of units cannot be recognized by the software.

Special Features Supporting the Retrieval of Physical Properties

Physical Units and Unit Conversion. Most physical quantities are associated with a unit serving as the standard measure for this entity. Even though much standardization has been done, there are still several different unit systems in use. Of course, this is also depending upon the area of application. In the Beilstein database most properties are measured in SI units but

---

1 The possibility for tolerance specification will be introduced in the Messenger software in early 1990.

there are still some quantities which are given in other units. In table IV the main units used in the Beilstein file are listed.

Table IV. List of Main Units used in Beilstein

Quantity	Unit	Unit Symbol
Amount of Substance	mole	mol
Angle	degree	deg
Concentration	percentage	%
	gram/liter	g/L
Density	gram/cubic centimeter	g/cm**3
Electric Potential	Volt	V
Energy	Joule/mole	J/mol
Entropy	Joule/mole*Kelvin	J/mol*K
Heat Capacity	Joule/mole*Kelvin	J/mol*K
Length	milli-, centimeter,...	mm, cm, ...
Mass	gram	g
Moment (Dipole)	Debye	D
Pressure	Torr	Torr
Temperature	degree Celsius	Cel
Time	second	s
Wavelength	nanometer	nm

To overcome the difficulty to remember all the units used in different numeric databases STN has developed a feature for the conversion of units. This enables the customer to work in his preferred set of units independent of the units used in the file. The Messenger software will automatically do the unit conversion and search or display the data in the appropriate units. The unit conversion capability enables the user:

- to specify units with numeric search terms
- to display the default unit for a property
- to set the system unit for a numeric property according to his/her convenience
- to select a common standard for the system units.

In particular, the customer may work in the default units, he may globally set units or he could overwrite a unit for a specific field. To illustrate this feature, a few examples are presented in Figure 8. In the first example, a search using the default units for the database is presented. If the customer formulates his query without specifying any units, then the default units are taken and no unit conversion is performed.

Consequently, the display of this property shows the property in the original unit. In the second example, the explicit specification of the unit in the search statement overrides the default unit of this field. The numeric range is converted from the default unit into the unit of this search field. A subsequent display of this property shows the original units again. In the third example, the unit for Heat Capacity is globally set by the Messenger SET command. In this case, the unit is changed for the rest of the session unless the customer changes it again or overrides it explicitly. All subsequent searches and displays are now presented in the new unit. As for the SEARCH command, the user could also override the unit for a display of a property. This is done in the DISPLAY command shown in the last example in Figure 8.

Retrieval of Missing Values. In large numeric databases like Beilstein or Gmelin (Handbook of Inorganic Chemistry), not all properties are available for each substance (see Section 3, 'The STN Implementation of the Beilstein Factual and Structure Database'). In other words, there are 'holes' in the file. This means that for a given substance, some properties are not present, either because they have not yet been measured, or because they were not known at the literature closing date for recording the respective substance. When a customer is searching for a property with specific values, it may be that he is missing potential hits because some property values are not in the database, although they could possibly overlap with the search range. In a normal numeric search these potential hits are not included in the answer set. However, in STN databases there is a possibility to search for these missing values or 'holes'. In Figure 9 there is an example for a search of the 'holes' in the field Melting Point (MP). This is done by ORing the numeric search with the term 'MP/FNA' (Field Not Available). It should be noted, however, that a simple search for missing values should not be performed because, in some cases, the number of holes is greater than the number of substances with values for this property and the system limits are easily exceeded. Thus, it is strongly recommended to use the 'FNA-Search' only if it is really necessary and combine the search always with some restrictions to keep the answer sets within the system limits.

Searching for Property Names. In addition to the numeric search capabilities for the properties, there is

```
Example: Default Units

=> s 120 - 140/cp
L1            3 120 J/MOL*K -  140 J/MOL*K /CP

Example: User Defined Units

=> s 0.05 - 0.08 kcal/mol*k /cp
L2            2 0.05 - 0.08 KCAL/MOL*K /CP

Example: Global Setting of Units

=> set unit cp=kj/mol*k
SET COMMAND COMPLETED

=> dis unit cp
CP                 DEFAULT:    J/MOL*K
CP                 CURRENT:    KJ/MOL*K

=> s 0.05 - 0.1  /cp
L3            2 0.05 KJ/MOL*K -  0.1 KJ/MOL*K /CP

=> d cp

L3  ANSWER 1 OF 2

Heat Capacity Cp:
Value (CP)                Temp.(CP.T)               Ref.  Note
(KJ/MOL*K)                (Cel)
-------------------------+-------------------------+-----+-----
0.12996                   26.7                      2     1
0.16513                   93.0                      2     1
0.06131 - 0.11887         -179.5 - -6.6             3     1
0.00314 - 0.12658         -258.1 - 26.9             5     1, 4

Reference(s):
2. Schlinger,Sage, Ind.Eng.Chem. 44<1952>2454,2456, CODEN: IECHAD
3. Todd,Parks, J.Amer.Chem.Soc. 58<1936>134, CODEN: JACSAT
5. Scott,Ferguson,Brickwedde, J.Res.Natl.Bur.Stand.(U.S.) 33<1944>4,
   CODEN: JRNBAG

Note(s):
1. Handbook Data
4. cp :beim Saettigungsdruck.
```

Figure 8. Examples for using the unit conversion capability

```
=> s 10 - 30/mp or mp/fna
             137 10 CEL -  30 CEL /MP
           38697 ALL/FA
           26568 MP/FA
           12129 MP/FNA
                  (ALL/FA NOT MP/FA)
L4         12266 10 CEL -  30 CEL /MP OR MP/FNA
```

Figure 9. Example for the retrieval of documents with missing values

also the possibility to search for the availability of the property itself. For all properties which are available for a given substance, the property name is indexed in the field FA (Field Availability) and in PH (Property Hierarchy). Those properties having only a literature reference but no numeric value in the database, have an index entry both in CT/CTM (Controlled Terms/Multi-Component System) and in PH. Hence, the three fields can be used for different purposes of availability searches:

- to retrieve a numeric value and/or a literature reference (Property Hierarchy: PH)
- to retrieve a numeric value (Field Availability:FA)
- to retrieve a literature reference only (Controlled Terms: CT/CTM).

Expanding the Field Availability provides the list of property names and the corresponding number of occurrences, i.e. the number of substances for which the property is available. The strategy to search for property names is very simple and it corresponds to searching controlled vocabulary in bibliographic databases. An example is given in Figure 10. Here, we have searched for the availability of numeric data for Molar Polarization.

If one is interested in either evaluated or non-evaluated property information, there is another possibility to search for the availability of numeric data using the keyword subfields ('xx.KW'). As described previously, the keywords 'HANDBOOK' and 'UNCHECKED' are indexed for each numeric entry in the corresponding keyword field. A search for these keywords together with the numeric values restricts the answer set to either evaluated or non-evaluated data satisfying the numeric query. If the query contains only the keyword term and not the numeric term, then this is equivalent to a search for the availability of this property restricted to either handbook or unchecked data. In the second example of Figure 10, an example for this type of search is listed. It should be noted that this kind of availability search cannot be done with the field FA.

Conclusions

The Beilstein database contains a comprehensive manifold of different physical property data. There are many numeric databases publicly available through online services but none of them can compete with Beilstein with respect to both the number of substances and properties. The implementation of the Beilstein database on STN has provided a number of new features and

```
Example: Availability of Data

=> s mpol/fa
L6          156 MPOL/FA

Example: Availability of Handbook Data

=> s handbook/mpol.kw
L7           64 HANDBOOK/MPOL.KW

=> d mpol

L7  ANSWER 1 OF 64

Molar Polarization:
Value (MPOL)              Ref.  Note
(cm**3/mol)
-------------------------+-----+-----
                          3     1, 2
                          3     1, 4

Reference(s):
3. Ebert, Eisenschitz, v. Hartel, Ph.Ch.<B> 1,110

Note(s):
1. Handbook Data
2. Molekularpolarisation im festen und im fluessigen Zustand.
4. Molekularpolarisation von Loesungen in Benzol und
   Tetrachlorkohlenstoff.
```

Figure 10. Search for the availability of physical properties

additional fields supporting, especially, the numeric entities. In addition, the Messenger software has been enhanced with new capabilities to provide a numeric data service within STN. Among these capabilities are a numeric range overlap detection, a concept for the specification of tolerances, a unit conversion feature, and a possibility to perform searches for missing values.

As shown in this paper, there are many possibilities to perform rather sophisticated searches of physical property information in the Beilstein database. Together with the features for the retrieval of substance and reaction information, the database supports the customer with different complementary choices of data access. With this respect, it can be stated that the Beilstein database may provide answers to almost any area of organic chemistry.

Acknowledgment

The funding of this work by the Federal German Ministry of Research and Technology is greatly acknowledged.

RECEIVED May 17, 1990

Chapter 9

Experiences of Two Academic Users

Evaluation of Applications in an Academic Environment

Gayle S. Baker[1] and David C. Baker[2]

[1]Engineering Library and [2]Department of Chemistry, University of Alabama, Tuscaloosa, AL 35487

Beilstein Online has been used and critically evaluated by both an engineering librarian and an organic chemist working in an academic environment. The database was found to provide ready access to a wealth of factual information on organic compounds. Search examples based on both text-term and substructures are provided. Positive features of the database include its coverage (acyclic and heterocyclic compounds back to 1830), and the abundance of critically evaluated data on organic compounds. Comparisons in content are drawn with other online databases.

The University of Alabama offers graduate and undergraduate degree programs in both the Department of Chemistry and in the Department of Chemical Engineering. Faculty and students in both programs have chemical information needs, much of which can be met by using the Beilstein Online database.

The authors of this paper, a faculty member specializing in synthetic organic chemistry and a reference librarian in an engineering library, are both experienced online searchers. The chemist uses Beilstein Online to solve problems in his own research laboratory, in which heterocyclic and carbohydrate compounds are synthesized and evaluated as antiviral and antitumor agents. The librarian uses the database to help chemical engineering students with chemical process design projects.

Use of *Beilstein*

Beilstein for many years has been a very important part of the synthetic organic chemistry program at the University of Alabama in that it provides a very efficient way to look for information on organic compounds, particularly intermediates in synthetic processes.

0097–6156/90/0436–0130$06.00/0

In academic circles, there seem to be two types of organic chemists: those who use *Beilstein* and those who do not. The group who do not use the handbook will cite two reasons for not using *Beilstein*: 1) it is always out-of-date, and 2) the language is German. While the last reason has, in years past, been generally regarded as an inexcusable one, the lack of foreign language skills among today's American students constitutes a serious barrier to using *Beilstein*. However, with the use of the *Beilstein Dictionary*, a compact, sixty-page booklet available from Springer-Verlag, the persistent worker, even with a very limited knowledge of German, can quite easily determine the principal physical properties of a compound or discern a process for its synthesis. While there is nowadays a trend in the use of English as the language for chemical literature, the majority of the literature of organic chemistry before ca. 1930 is certainly in German (*1*), and the use of *Beilstein* goes hand in hand with accessing that important segment of the chemical literature.

The other reason posed for avoiding the use of *Beilstein* is that it is not up-to-date. That drawback is rapidly being removed. The Beilstein Institute has embarked upon a major automation project in order to produce an online database which shall essentially be caught up with the current literature within the next few years. (Refer to the chapter in this volume by C. Jochum.)

Database Description

Currently, Beilstein Online, as it is implemented on STN, contains information from volumes 17—27 (Main Series and Supplementary Series I—IV), covering the heterocyclic compounds. This is critically evaluated data on approximately 350,000 records from the year 1830 through 1959. In addition, there are excerpts from the Supplementary Series V for the period 1960 through 1979, covering over 1.2 million compounds, much of which is unchecked and not yet in the printed version of *Beilstein*. These heterocyclic compounds include many substances important to medicinal and natural products chemistry, as well as to agricultural chemistry, and related disciplines. Very recently, volumes 1—4, the acyclic compounds, some 110,000 substances with critically evaluated data from 1830—1959 (Main Series, Supplementary Series I—IV), were added to Beilstein Online.

It is important to note, from the standpoint of the organic chemist, that ca. 1960 marks the advent of nuclear magnetic resonance spectroscopy as a regularly used tool for identification of organic compounds. This powerful spectroscopic technique has literally revolutionized the chemist's ability to determine the structure of organic compounds. Hence, one will find that the large amount of spectroscopic data excerpted and referenced in records from this Supplemental Series V (much of which is not yet in print) is a very valuable resource for the practicing organic chemist. In addition, the older spectroscopic techniques—infrared spectroscopy and ultraviolet spectroscopy—are well cited from 1959 retrospectively to ca. 1930.

Each substance in the Beilstein database has a unique Beilstein Registry Number (BRN). Other information included in its associated record are the chemical substance name, chemical structure, and molecular formula and weight. There are optional fields available for physical and chemical properties, chemical reactions, and preparations. (See Figure 1 for a hierarchical layout of the Beilstein record.) Each record includes a data field, called Field Availability, which can be used to

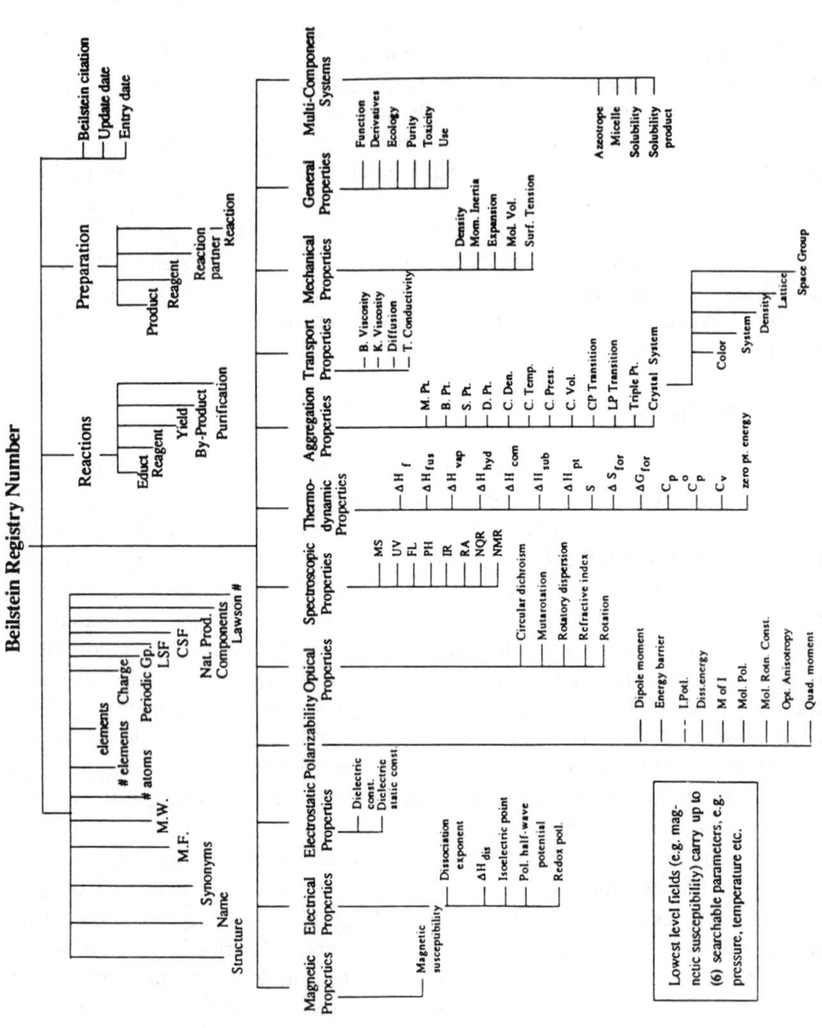

Figure 1. Structure of Beilstein Online Record. (Reproduced with permission from Ref. 2. Copyright 1989 Springer–Verlag.)

determine the availability of optional fields. There are over 200 possible searchable and/or displayable fields, and a majority of these may occur more than once in a single record.

Records are retrieved by searching data fields for factual information, for example, a molecular formula and a range of boiling points. Structure queries may also be constructed to perform exact or substructure searches. One may create the query online using the STN Messenger structure-building commands. These queries may also be created offline using software such as MOLKICK or STN Express, either of which allows one to draw a molecular structure and then generate the appropriate structure-building commands for uploading while one is online with the Beilstein file. A combination of the two methods may also be used. Terminal emulation software for the PC, such as PC-PLOT, can be used to display structures from Beilstein Online records as they might appear on a true graphics terminal.

Examples of Searches

Carbohydrate chemists are very much concerned with the optical rotation [termed optical rotatory power (ORP) in Beilstein Online] of their compounds, which is a physicochemical datum as important as the melting point, boiling point, or other physical constant. Suppose that one has chemical data for an unknown sugar which shows that its formula is $C_{12}H_{22}O_{11}$ and that it has a measured optical rotation of $+67^{\circ}$. The first step of a search for information on Beilstein Online retrieves 333 items for records having the molecular formula, $C_{12}H_{22}O_{11}$. Since the optical rotation, as it is normally measured, exists plus or minus ca. 3 degrees of a nominal value, the optical rotation field should be searched for a range of values. The second step, introduction of a range for ORP of 65—70°, narrows down the first step to those records with an optical rotation with a range of 65—70° and retains only 13 records. (See Figure 2.) The output from one of the records is printed using the default option of the "display" command, which includes the identification data fields (Beilstein Registry Number, Molecular Formula, Compound Name, Synonym, Formula Weight, Beilstein Source Identification number, Lawson Number, Structure) and any displayable fields used in the search terms. The output provides pertinent references to the literature; however, in many cases the information as provided online is sufficient.

In a more typical example, an intermediate compound in the preparation of some antitumor substances, we needed to know something about the preparation of some substituted pyridines, especially 2-amino-3-iodopyridine. Since our experience had shown that searching by chemical name quite often yielded a set of zero records (due to the difficulty in constructing the *exact* chemical name), we opted to search for records having a molecular formula of $C_5H_5IN_2$ and the chemical name segment of "PYRIDIN," the German spelling of pyridine. (See Figure 3. Note that it has since been pointed out to us that a better strategy would have been to search for "PYRIDIN#", as the latter would have also covered the English "PYRIDINE". The SY-field includes English terms if indeed the author originally wrote in English.) The first record in the resulting set of two records was 2-amino-3-iodopyridine. To see what data is available, we used the "trial" option of the "display" command, which prints the contents of the Field Availability Field, a list of displayable data

```
=> S C12H22O11/MF
L1          333 C12H22O11/MF

=> S L1 AND 65-70/ORP
               2235 65-70/ORP
L2                13 L1 AND 65-70/ORP

=> D L2 1-13
```

(Displayed L2 1-13 and examined data presented for "ORP")

***Found** pertinent data in ans. 13:*

```
L2 ANSWER 13 OF 13

BRN  90825  Beilstein
MF     ***C12 H22 O11***
CN   .beta.-D-fructofuranosyl-.alpha.-D-glucopyranoside
     .beta.-D-Fructofuranosyl-.alpha.-D-glucopyranosid
SY   Saccharose
FW   342.30
SO   4-17-00-03786; 5-17
LN   17647; 17644
```

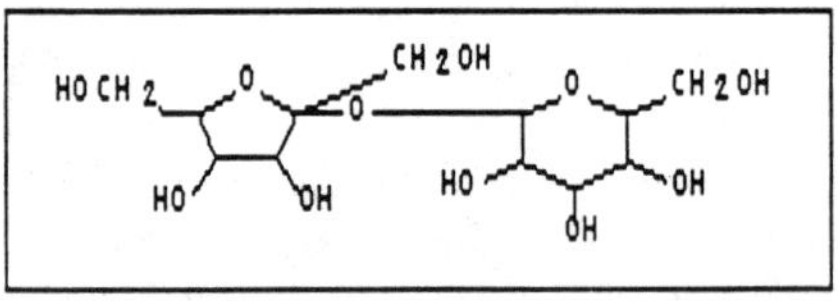

Figure 2. Search for a Sugar with MF $C_{12}H_{22}O_{11}$ and Optical Rotation = +67°.

```
=> S C5H5IN2/MF AND PYRIDIN
          11 C5H5IN2/MF
       23447 PYRIDIN
L2         2 C5H5IN2/MF AND PYRIDIN

=> D L2 1-2

L2 ANSWER 1 OF 2

BRN  471506  Beilstein
MF     ***C5 H5 I N2***
SY     ***2-Amino-3-iod-pyridin***
FW   220.01
SO   5-22
LN   27379

     N        NH2
  . :  .
C.    : C.
:       .
:       .
:       .
C .   : C .
  . C:      I

L2 ANSWER 2 OF 2

BRN  471425  Beilstein
MF     ***C5 H5 I N2***
SY     ***2-Amino-6-iod-pyridin***
FW   220.01
SO   5-22
LN   27379

I.      .N:      NH2
 .  .    :  .
   .C.      :C.
    :        .
    :        .
    C.      :C
      .   :
       .C:
```

Figure 3. Search for 2-Amino-3-iodopyridine.

fields which are present in the record. (See Figure 4.) Preparation data are printed by indicating the "PRE" field designation with the "display" command. Among the interesting information contained in the references are the journal CODEN, a unique code assigned to journal titles and useful in document retrieval. This record also contains a reference to the abstract of one cited article in *Chemical Abstracts*, a feature that is especially welcome whenever the bibliographic source is an uncommon journal and one would like to consult the abstract in *Chemical Abstracts*.

The isolation of caffeine is a procedure which is still performed as an academic laboratory exercise, and it provides an example of a compound where a tremendous amount of data is available. (See Figure 5.) Once the appropriate record is found by searching for the molecular formula, the "trial" option of the display command is then used to generate a listing (at no charge) of fields which are available for the record, as well as the number of multiple occurrences for each field. The "trial" display of the Beilstein Online record for caffeine indicates, for example, that there are 71 different references for "Preparation" and 33 references for "Isolation from Natural Product". Some other fields having multiple occurrences are Melting Point (44 refs.), Fluorescent Spectrum (2 refs.), Solubility (75 refs.), and Biological Function (2 refs.). If we had used the "all" option with the "display" command for the caffeine record, we would have generated a large amount of text as well as a sizeable bill. It is important to point out that we advocate the use of the "trial" option with Beilstein Online to determine the amount of data available in a record. Otherwise, one might be quite surprised with the length of the output and the amount of the connect-hour charge. This display is also useful to the novice user of Beilstein Online as it supplies the display codes of the available fields.

MOLKICK

MOLKICK is a memory-resident program that runs on an IBM PC or compatible microcomputer and facilitates quick and easy generation of chemical structures. One simply uses a mouse to draw structures and select menu options. STN structure-building commands are generated from a structure using the F9 key. The commands are stored in a file which can be uploaded during a search of Beilstein Online on STN to create a structure query. (See Figure 6 for an example.)

Among the advantages of creating a structure using MOLKICK are that one can draw a structure in a free-hand manner and that one does not need to know the STN Messenger structure-building commands. In addition, there is a savings in connect-hour charges by generating the commands offline prior to beginning a search session.

Comparison of Beilstein Online with Other Factual Databases

Two online databases that contain data fields common with those available on Beilstein Online are Heilbron (an online version of Heilbron's *Dictionary of Organic Compounds*, 5th edit. and *Dictionary of Organometallic Compounds*) and the Merck Index Online. [Information for these files was taken from the Dialog database descriptions for files 303 (Heilbron, December 1986) and 304 (Merck Index Online, January 1989).] All three are oriented around chemical compounds and are thus

```
=> D 1 TRIAL

L2  ANSWER 1 OF 2

MF      ***C5 H5 I N2***

     Code    Field Name                              Occur.
     -------+----------------------------------+------
     MF      Molecular Formula                       1
     SY      Synonym                                 1
     FW      Formula Weight                          1
     SO      Beilstein Citation                      1
     LN      Lawson Number                           1
     PRE     Preparation                             2
     MP      Melting Point                           2

=> D 1 PRE

L2  ANSWER 1 OF 2

Preparation:
PRE
     Reference(s):
     1. Azev et al., Chem.Heterocycl.Compd. (Engl.Transl.), 10, <1974>, 687, CODEN:
        CHCCAL

PRE
     Reference(s):
     1. Kowtunowskaja-Lewschina, Tr.Ukr.Inst.Eksp.Endokrinol., 18, <1961>, 350,352,
        CODEN: TUEEA9
        CA: 7921, 58, 1963
```

Figure 4. References for the Preparation of 2-Amino-3-iodopyridine.

MF ***$C_8 H_{10} N_4 O_2$***

Code	Field Name	Occurrence
MF	Molecular Formula	1
CN	Chemical Name	1
SY	Synonym	1
FW	Formula Weight	1
SO	Beilstein Citation	1
LN	Lawson Number	1
PRE	Preparation	71
MP	Melting Point	44
SP	Sublimation Point	2
REA	Chemical Reaction	74
CTGEN	Structural Data	1
RSTR	Related Structure	10
INP	Isolation from Natural Product	33
PUR	Purification	1
CTSKC	Skeletal Characteristics	2
DM	Dipole Moment	2
IP	Ionization Potential	3
CPD	Crystal Property Description	1
CTCRY	Crystal Phase	2
CTP	Crystal Transition Point	1
CTLIQ	Liquid Phase	2
CSYS	Crystal System	1
DEN	Density (crystal)	2
DEN	Density (liquid)	2
CTCAL	Calorific Data	2
CPO	Heat Capacity CpO	1
RI	Refractive Index	2
CTOPT	Optics	2
NMRS	NMR Spectrum	1
IRS	Infrared Spectrum	7
CTESP	Electronic Spectrum	1
EAS	Electronic Absorption Spectrum	19
EAM	Electronic Absorption Maximum	3
FLS	Fluorescence Spectrum	2
CTOSM	Other Spectroscopic Methods	1
CTMS	Mass Spectrum	4
DIC	Dielectric Constant	1
CTELE	Electrical Data	3
CTECB	Electrochemical Behavior	12
DE	Dissociation Exponent	12
SLB	Solubility	75
CTSOLM	Solution Behavior	11
CTLLSM	Liquid/Liquid Systems	25
CTLSSM	Liquid/Solid Systems	7
CTLVSM	Liquid/Vapor Systems	1
CTGASM	Gas Phase Behavior	1
CTTRAM	Transport Phenomena	2
CTENEM	Energy of MCS	2
CTBSPM	Boundary Surface Phenomena	3
CTASSM	Association	20
BF	Biological Function	2
CDER	Chemical Derivative	1
CTUNCH	Unchecked Data	14

Figure 5. "D Trial" for Caffeine.

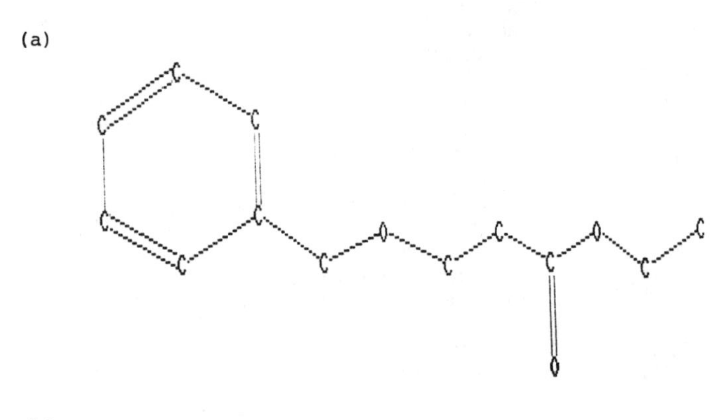

(b)

```
set pagelength scroll
STR

SET BOND SE
GRA C6,6 1,1 C6,11 C3
NOD 8 O,13 O,12 O
BON 11-12 DE
BON 1-2 1-6 6-5 5-4 4-3 3-2 N
CON 15 12 E1
CON 6 5 4 3 2 7 8 9 10 E2
CON 13 14 E2
CON 1 11 E3
END
set pagelength 33
```

Figure 6. (a) MOLKICK Structure and (b) CAS Upload File Generated from the Structure.

factual databases, i.e., they are databases "containing substances (chemicals) and the data associated or connected with the substances" (*3*). On the other hand, the CAS ONLINE CA File is by comparison largely a bibliographic database, with a relatively small amount of factual information being available in the Registry File.

All three of the databases have bibliographic references to original literature, including patents, in the majority of records. It is worth noting that only the last name of an author is provided on Beilstein Online, while Heilbron and the Merck Index Online files include author's initials.

Each of these databases contains a number of searchable and displayable fields associated with physicochemical properties of chemical compounds. One may search a range of values in many of the fields, such as the Melting Point and the Boiling Point. In addition, each database provides a means for the searcher to determine which of the optional data fields are actually present in a record: Field Availability in Beilstein Online, Data Tags and Reference Tags in the online Heilbron, and References Present and Data Present in the Merck Index Online. All provide free display formats for these "data availability" fields.

Beilstein Online is by far the most comprehensive of the three in the number of possible searchable and/or displayable data fields. It has over 200 fields and subfields. The Heilbron and Merck Index Online files have only 33 and 41 fields, respectively.

While Heilbron and Merck Index Online are far less comprehensive than Beilstein Online, much commonly needed data, however, are quickly retrieved from these latter two sources. Beilstein Online currently has some 460,000 complete records with 1.2 million excerpts. Heilbron, by comparison, covers 175,000 chemical compounds (organic and organometallic) contained in 70,000 records with references. The Merck Index Online file (10th edition) is limited to 30,000 substances, principally pharmaceuticals and biologicals, in 10,300 records.

Beilstein Online Search Costs on STN

The connect-hour fee for Beilstein Online on STN, which is currently $49.00 per hour, seems very reasonable. (See Figure 7) The fact that there is a charge of $2.70 per search term indicates that the Beilstein Institute is following the pricing policies of CAS (*4*). Substructure searching is a bargain at about half the charges for a substructure search carried out on CAS ONLINE.

The price of an offline print is the same as an online display, $1.60 per field and answer. If one is indiscriminate in using the "all" option of display, a sizeable charge can be quickly accumulated. This underscores the importance of using the free "trial" option of display to ascertain the amount of data present in a record.

Special discounts are available. Academic users whose libraries subscribe to the *Beilstein Handbook* are entitled to a 50% discount on all charges. In addition, there is a special 90% discount for classroom use of the file. There is a 10% discount to all other subscribers of the printed handbook.

Advantages of Using Beilstein Online

The principal advantage of Beilstein Online is the coverage, essentially 1830—1979,

Beilstein File	**U. S. Dollars**
Connect Hour Fee (per hour)	$ 49.00
Search Term, per term	2.70
Search	
Substructure Search	
Full File	49.00
Sample	FREE
Exact Structure Search	
Full File	9.70
Range	9.70
Sample	FREE
Family Search	
Full File	9.70
Range	9.70
Sample	FREE
Online Display/Offline	
Print (per field and answer)	1.60
Display Field, each	1.60
Display Fields/Formats	
BRN, MF, FA, CN, SY, FW, SCF, LSF, NTE, OCC, TRIAL	FREE
Default Format (dynamic): IDE plus HIT	
IDE+	4.80+
HIT (per field)	1.60

Special Subscription Prices:

50% discount to educational buyers of the *Beilstein Handbook*

90% discount for special classroom use

10% discount to all other regular buyers of the *Beilstein Handbook*

Figure 7. STN International Fees and Prices (Effective January 1, 1989).

covering volumes 17—27 and volumes 1—4, as discussed earlier. Costs are cheaper than CAS ONLINE if one is judicious in the display of data fields. The structure searching and the daytime connect time charges are certainly less expensive.

Chemists and chemical engineers are frequently concerned with preparative information along with the starting material, i.e., the "educt" in Beilstein terminology. Preparative routes for a compound can be searched by educt. While CAS ONLINE now offers a CASREACT database, it is extremely limited, having at present only some 39,000 records from about 100 journal titles from 1985 to the present. (This information was taken from the STN database description sheet for CASREACT.)

As mentioned earlier, Beilstein Online has more data fields than either the Heilbron database or the Merck Index Online database.

The inclusion of the journal CODEN in bibliographic references is helpful in the retrieval of information in sources that may not be owned by one's library. Also, the listing of the *Chemical Abstracts* reference is very useful.

The MOLKICK software simplifies structure searching by allowing almost "freehand" drawing of a chemical structure. The program translates the structure into appropriate commands to upload for building a structure for structure searching on STN.

Limitations of Beilstein Online

There are some limitations of the Beilstein Online system as it currently exists on STN.

We have discovered that structures of some complex compounds (e.g., adenosine and related nucleosides) may not be retrieved online to be modified and used in structure searches. Unless one has access to offline structure-building software like MOLKICK, the cost of connect time to build a complex structure query online using the STN structure-building commands can become quite high.

Author searching can be limited due to the fact that Beilstein Online contains only last names. A search for common names like "Smith" or "Schmidt" can produce a very large set. Unless a name is quite unique, an author search should be combined with a search for one or more other terms.

The large amount of data which is available for common compounds can cause "information overload". This is especially true for the novice user who is unfamiliar with the structure and data fields of the Beilstein Online records. It is recommended that one first use the "trial" option for the display command to screen available data, then proceed by selecting only the display fields of interest.

Both the Heilbron and the Merck Index Online databases have CAS Registry Number fields; however, Beilstein Online does not. This feature would facilitate many searches by providing an excellent means for linking searches with information gleaned from CAS ONLINE and vice versa.

While direct file crossover from CAS ONLINE to Beilstein Online is allowed on the STN search system, STN issues a separate academic account for Beilstein Online. This account cannot access CAS ONLINE at the academic rate. Thus file crossover searching can be prohibitively expensive for an academic user.

Both Heilbron and the Merck Index Online databases are available on Dialog's

Knowledge Index after-hours search system. The cost is a flat $24 per hour, with no additional search term or field display costs, which may be an advantage for a non-academic user or an academic user whose library does not subscribe to the printed counterparts of these files.

Finally, there are no data available after 1979 on Beilstein Online at the present time.

Summary

Overall, we are impressed with Beilstein Online. It offers the most comprehensive, online source of *critically reviewed* information on organic compounds. The database extends coverage on selected sections of *Beilstein* back to 1830. As this coverage develops, the database will become a formidable source for physicochemical data, spectroscopic data, and information on reaction processes for both the chemist and chemical engineer.

Acknowledgments

The authors thank Robert C. Badger of Springer-Verlag (New York) for making available to the authors some free connect time on Beilstein Online. Carolyn Rhyne is especially thanked for transcribing D. C. B.'s taped address (*5*) for use in writing this chapter.

Literature Cited

1. Ash, J. E.; Chubb, P. A.; Ward, S. E.; Welford, S. M.; Willett, P. *Communication, Storage and Retrieval of Chemical Information*; Ellis Horwood: West Sussex, U.K., 1985; p. 44.
2. Heller, S. R.; Milne, G. W. A. *Online Searching on STN: Beilstein Reference Manual*; Springer-Verlag: New York, 1989; pp. 1—6.
3. Heller, S. R. *Database* **1987**, *10* (4), 50.
4. Pemberton, J. K. *Online* **1988**, *12* (2), 7.
5. Baker, D. C.; Baker, G. S. *Abstracts of Papers*, 198th National Meeting of the American Chemical Society, Miami, FL; American Chemical Society: Washington, DC, 1989; Abstract COMP 5.

RECEIVED May 17, 1990

Chapter 10

The Lawson Similarity Number (LN)

Offline Generation and Online Use

Alexander J. Lawson

Beilstein Institute, Varrentrappstrasse 40–42, D-6000, Frankfurt am Main, 90, Federal Republic of Germany

The Beilstein Database (at time of writing) contains many individual properties of ca. 3 million compounds, principally numerical data and preparative methods.

This information can be retrieved directly by value (e.g. pK_a between 2.0 and 2.5), by type (e.g. all substrates where the dipole moment was measured using the Stark Effect), from structural information (substructure search, molecular formula, nomenclature), or any combination of these methods.

This combination of retrieval tools is very powerful and can rapidly extract the maximum amount of information from this enormous database, since the searcher can design his query with a precision unavailable on other chemical databases till now.

Sometimes, however, a controlled relaxation of this precision is desirable and advantageous : the experienced chemist knows that many types of information can be transferred from the particular reported case to other analogous cases. For instance, preparative methods are often equally applicable to classes of compounds, and certain physical data can be used as a starting point for the estimation of unreported data in the same class of compounds.

This is the reason why the Beilstein Database is equipped with a retrieval tool for classification of structures, the so-called Lawson Number (LN). This number enables the searcher to investigate a family of structures, in particular the family referred to as **"positional isomers"**. In general it is not easy, cheap

0097–6156/90/0436–0143$06.00/0

or quick to retrieve positional isomers with traditional substructure searching, and the LN can be regarded as an additional help to existing tools, **not** as a substitute.

The LN is a two-byte unsigned integer, which means that in principle the number can take any value in the range 0 - 65535.

To date, the user is only familiar with values in the range 9-32759, since the implementation of the LN was planned to be carried out in two phases, one of which is already completed. It is the purpose of this paper to outline the nature of the LN and its use, including a description of the second implentation phase in the Beilstein database.

The First Implementation Phase

The first implementation of the LN was active in the Beilstein database from its inception (on STN) in December 1987 to mid 1990. The general concept and examples of LN searches have been published elsewhere [1,2]. The following points are relevant to the present discussion :

a) As noted above, the range of LN values actually used in practice in the first phase was 9-32759, (i.e. ca. one half of the full range available).

b) Any particular compound may possess several LN, one for each "chemically significant" fragment in the molecule. In general, most compounds have 2-3 LN values. An analysis of 1.8 million compounds of the Beilstein database gave the following picture :

25.1 % have 1 LN value
39.4 % have 2 LN values
24.0 % have 3 LN values
8.5 % have 4 LN values
3.0 % have >4 LN values

Thus the LN is an economical (but non-unique) description of structural properties, since each structure is characterized by less than 5 bytes on average.

c) The spread of LN values is very even ; the occurrences of the following 3 small ranges (1000-1009, 19000-19009, 27000-27009) are quite typical :

	LN	Cmpds.	LN	Cmpds.	LN	Cmpds.
	1000	86	19000	67	27000	60
	1001	40	19001	35	27001	49
	1002	57	19002	28	27002	181
	1003	16	19003	26	27003	200
	1004	3	19004	19	27004	81
	1005	17	19005	119	27005	39
	1006	96	19006	121	27006	45
	1007	166	19007	98	27007	134
	1008	81	19008	44	27008	44
	1009	120	19009	19	27009	71
Average :		68		57		90

Thus the average LN occurrs in ca. 70 compounds in a database of 1.8 million compounds, equivalent to 0.004 % (with several notable exceptions, such as MeO (LN=289), which occurs in one compound in seven).

This results in quick response times.

d) The retrieval quality of the LN was quite satisfactory, according to the reports of expert users . In particular, when properly understood the LN was capable of yielding similarity searches of a certain type (structure browsing, as in the printed version of the Beilstein Handbook), and could even be used with certain types of generic queries [3].

The following is a good example of the power of the LN, performed at a point of time where the database corresponded to 350,000 heterocyclic compounds :

The search strategy was designed to retrieve all dihalogenated pyridine carboxylic acids.

The query was formulated as :

The names of the 13 hits obtained are shown in Fig. 1

e) However, the LN also had drawbacks in the first phase. These are listed again here although they were clearly indicated in earlier publications [1,2] :

5,6-dichloro-nicotinic acid

2,6-dibromo-isonicotinic acid

2,6-dichloro-isonicotinic acid

3,5-diiodo-isonicotinic acid

3,5-dichloro-isonicotinic acid

4,6-dichloro-nicotinic acid

2,6-dichloro-nicotinic acid

3,5-dibromo-pyridine-2-carboxylic acid

3,5-dichloro-pyridine-2-carboxylic acid

2,6-diiodo-isonicotinic acid

5-chloro-4-iodo-pyridine-2-carboxylic acid

4,5-dichloro-pyridine-2-carboxylic acid

4,6-dichloro-pyridine-2-carboxylic acid

Fig. 1

Hits obtained by the use of the STN search :
s 26334/LN and C=6 AND N=1 AND O=2 AND X=2

1) The LN could not be simply generated by the user at query-time. There was no mechanism to arrive at a particular set of values for a particular structure, apart from finding that structure ONLINE, and then using the LNs as search terms to widen the view around the given structure.

2) It was not clear to the user under what exact conditions he was searching when using the LN term. This is to some extent a natural consequence of the very act of browsing (if you know **exactly** what you are looking for then you are not browsing!), but there was still a need for a clearer definition.

3) As a direct result of the point 2), in certain cases there were inevitable clashes between the subjective view of similarity and the hit lists returned. In particular, the user was sometimes confronted with a spattering of ring systems which overstepped his subjective view of chemical similarity. (For instance, a query based on pyridines might possibly contain a few methyl pyrroles among the otherwise sensible hits, which would then be considered as false drops).

A second example will illustrate this point nicely :

The search strategy was directed at the class of N-methyl X-benzoyl Y-phenyl piperidines, in which the opositional locants X and Y could vary.

The specific structure **1** (Fig.2) was found in the database by other means (point **e** 1, above).

The LNs for this compound were then used for the search, combined with the molecular formula , i.e. :

S 25555/LN AND 2817/LN AND C19H21NO/MF

The use of this query on the BEILSTEIN database resulted in the hits shown in Figure 2.

The hitlist was returned practically instantaneously, and the search is inexpensive.

However ..

.. hits **1** **2** and **3** are positional isomers, and are what was wanted. Hits **4** and **5** on the other hand, although positional isomers of each other, are not what was required, since the nature of the ring is not similar enough for the chemist (point **e** 3, above).

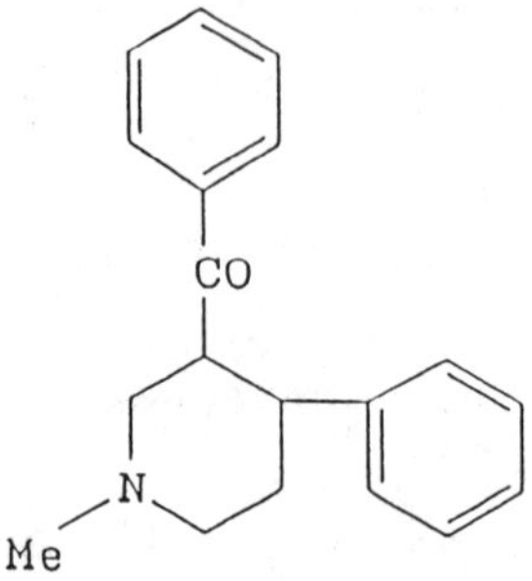

1

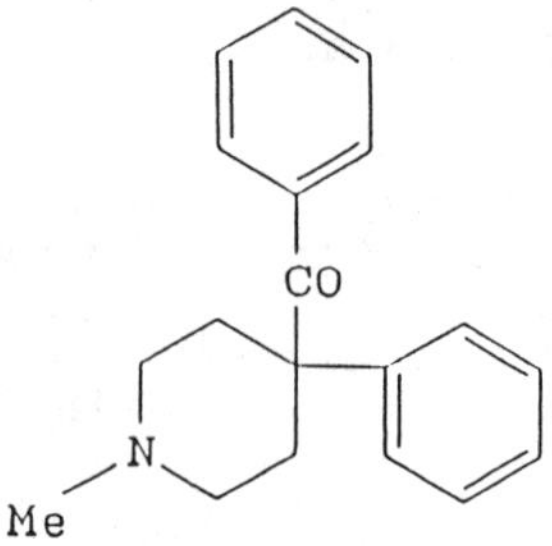

2

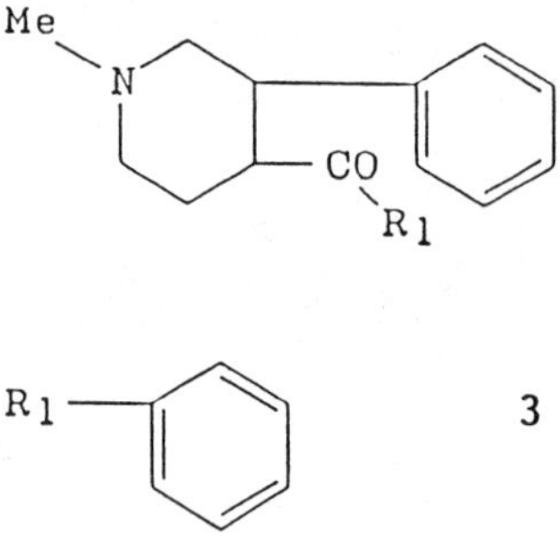

3

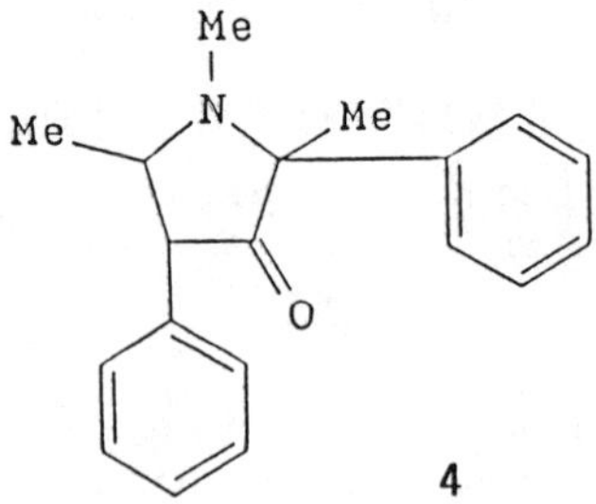

4

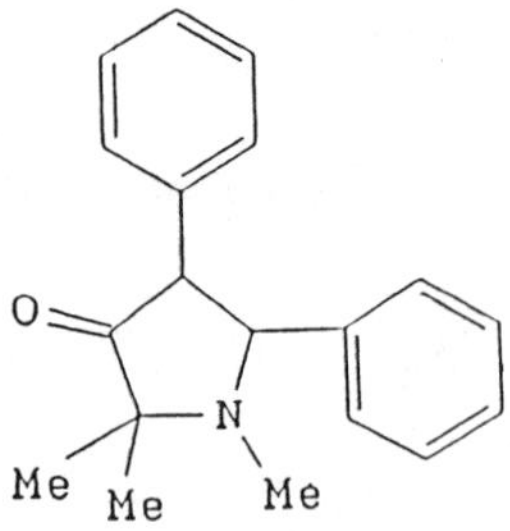

5

Fig. 2

Hits obtained by the STN search : S 25555/LN AND C19H21NO/MF

Note that these can sometimes be filtered out by nomenclature terms, but nomenclature **alone** is rarely capable of retrieving positional isomers in a dependable fashion, because of the non-systematic usage encountered in even the best of databases. For instance, not all the hits of Fig. 1 necessarily contain the nomenclature term "pyridine" , since "nicotinic acid" and "isonicotinic acid" are perfectly acceptable names. Similarly, the hits of Fig. 2 are not necessarily named using the term "benzoyl", or even "piperidine". In fact, the names found were based on "phenyl piperid-X-yl ketone".

This was the experience gained in the first implementation phase. The strengths and limitations of the approach were now clear, from the point of view of the user.

The aim of the second phase is to remove the remaining drawbacks to the use of the LN. These are now discussed in the next section.

The Second Implementation Phase

The second phase will be implemented in mid-1990.

In the course of the second phase, the drawbacks listed in points e 1, 2, and 3 on page 147 will be counteracted in the following manner.

To point 1) :

The calculation of the LNs for any given structure can be performed fully automatically and rapidly by PC-software, offline, in preparation for an Online session. The software compiles the query , in ASCII format, which can be simply uploaded at session time. The structure is drawn using a mouse, and the query is presented in the proper form for either STN, DIALOG or MAXWELL ONLINE (ORBIT/BRS).

The following example illustrates this point (in this case, but not exclusively, in the environment of the SANDRA package, from Version 3.0 onwards).

(Insert Fig.3)

To point 2:

The user is often unclear about exactly what is being searched in an LN query. In point of fact, the LN is a flexible tool with many facets and possibilities. For the present purposes, it is sufficient to concentrate on one particular aspect of the use of the LN, namely **in connection with the full molecular formula** of any given

S 25555/LN AND 2817/LN AND C19H21NO/MF

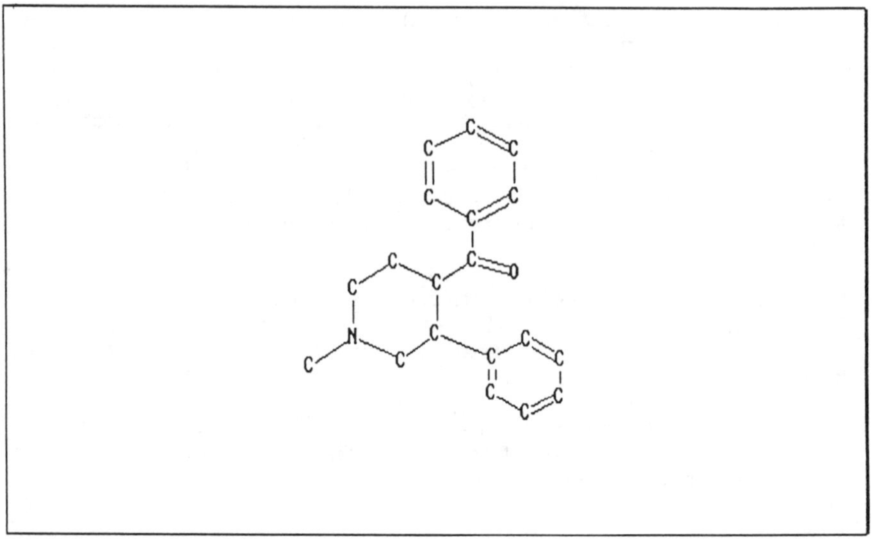

Save Return

Fig. 3

Automatic Generation of LN Search Command (STN) Offline, directly from the structure input on PC-Software

compound. In this case, the hitlist obtained can be precisely and simply defined (also for the first phase), as follows :

The contents of the hit-list is :

that set of molecules generated by (possibly repeated) movement of any group of atoms **from** any branching C-atom **to** any other non-terminal C-Atom, without crossing the natural boundary of an exoyclic heteroatom, coupled with generally free relocation of C-C multiple bonds (if present).

This sounds complex, but an example will illustrate what is meant:

Example : 2-tert-butoxy-hexane has the following values:

LNs : 335 ; 318
MF : C10H22O

The search is carried out on the logic :

S 335/LN and 318/LN and C10H22O/MF

and the hit-list would contain (in this case) :

2-tert-butoxy-hexane
3-tert-butoxy-hexane
1-tert-butoxy-2-methyl-pentane
1-tert-butoxy-3-methyl-pentane
1-tert-butoxy-4-methyl-pentane

(see Fig. 4. I have taken considerable liberties with IUPAC nomenclature to make it clear here what is happening.)

In other words, the LN "recognises" two movable groups in the original molecule, a secondary methyl in the hexane and a tert-butoxy, and returns the combinations found. None of the methyls in the tert-butoxy are movable in the original molecule since this would involve a change in the total amount of branching).

This is the general definition, and in the first phase this also allowed (as a natural consequence) the inclusion of other ring types **when a ring bond is moved.** In the second phase, this particular degree of freedom (generally regarded as stretching the positional isomer concept too far, see point **e** 3 above) will be counteracted by the introduction of specific shape discrimination, which will now be discussed, below.

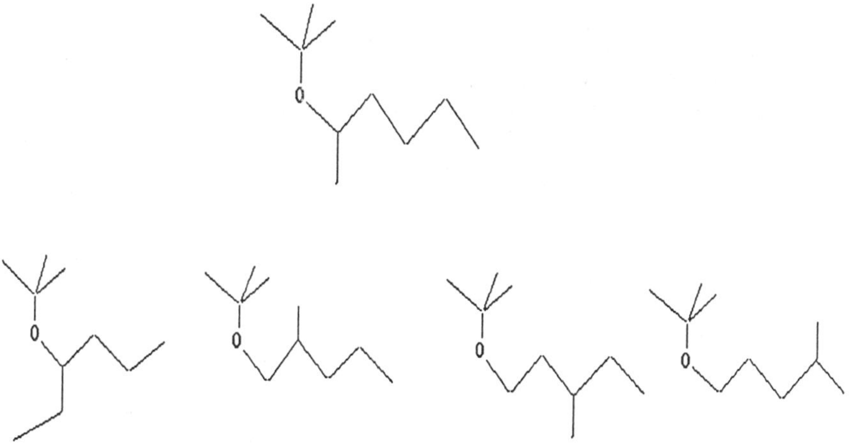

Fig. 4

The Theoretical Hit list from the LN Search :
S 335/LN and 318/LN and C10H22O/MF

To point 3 :

From the above discussion, it is clear that certain common types of shape should be explicitly recognised by the LN analysis. This will be implemented in the second phase as follows :

positions of these heteroatoms.

- all mono and bicyclic systems with 18 or more ring-atoms will be grouped together without explicit registration of the ring size, or ring positions of the heteroatoms.

- for tricyclic (and greater) ring systems, there will be a default assignment of a single LN to denote this status. Further discrimination could also be based on "present" or not- present" in a standard small library of shapes (for common shapes, such as steroid skeletons, morphane, adamantane etc.).

The important features of this shape discrimination, as far as the user is concerned, are as follows :

i) The assignment is transparent, inasmuch as it is possible to list the structures and their numbers in a printed reference work.

ii) However, this is not necesary. The generation of the LNs for shapes is completely automatic in the PC-software mentioned above.

iii) The LNs responsible for shape discrimination are easily distinguishable from the LNs for chemical functionality, degree of unsaturation, Carbon-number etc., since they have values greater than the 15-bit limit (32767). They contain information which is **supplementary** to the information of the LNs in the lower range, and may be used (or otherwise) in any particular search, completely at the discretion of the user.

To make this point clearer, consider again the query of Fig. 3 and the hit list of Fig.2.

After full implementation in the second phase, this query would be extended to :

S 32841/LN AND 25555/LN AND 2817/LN AND C19H21NO/MF

A set of ca. 8000 shape systems will be defined and allocated a value of the LN in the range 32776-40951.

These will be precisely defined according to the following scheme :

As a general principle, the extent and position of unsaturation (double, triple bonds) in the shape systems will not be taken into account at all. Only ring **skeletal shapes** will be registered, on the following basis :

- all mono and bicyclic systems (including fused, bridged and spiro) of 3-10 ring-atoms, containing 0, 1 or 2 heteroatoms of the set **(O,N,S)**, in any combination, in any ring position will be uniquely recognised and assigned a unique LN

Comment This covers the large majority of commonly encountered rings, such as furan, the penicillin-type skeleton (6-thia-1-aza-[3.2.0]bicycloheptane) etc. Here the user is generally interested in hits with precisely the given skeleton.

- all mono and bicyclic systems (including fused, bridged and spiro) of 11-17 ring-atoms, containing 0, 1 or 2 heteroatoms of the set **(O,N,S)**, in any combination, but without explicit discrimination of the ring positions of the heteroatoms, will be recognised and assigned an LN.

Comment In ring systems of this size, the reported occurrences are relatively low, and hence a less precise definition (here the relative location of heteroatoms) is acceptable in the context of similarity, especially when one takes into account the inherent geometrical flexibility of these systems.

- all mono and bicyclic systems containing more than two heteroatoms and/or heteroatoms other than N, O, and S will be grouped into special classes, which could be generated by the PC program offline. (Incidentally, each separate LN of a particular structure is identified explicitly and visually by the PC program with the corresponding fragment).

Now the hit list would contain only hits **1** to **3**. The hits **4** and **5**
would be retrieved by the query :

S 32818/LN AND 25555/LN AND 2817/LN AND C19H21NO/MF

iv) The LN shape discrimination is based on a pragmatic view of similarity, but does not compete with alternative, absolute views of ring structure (including multiple bonding, points of attachment, embedded fusion, etc.). These can be dealt with in a completely general manner using substructure searching, or an all-embracing ring-coding algorithmus. The LN does not attempt this, but rather concentates on a **classification**, in an attempt to bring similar entitites together in a chemically meaningful way, based solely on the input of a typical member of the chemical class.

This, of course has long been a goal of organic nomenclature, which can only be partially fulfilled, for a number of reasons (see above). Nevertheless, the value of nomenclature used in combination with the LN should not be underestimated. The next and final section of this paper touches on the use of nomenclature based on the same general philosophy which has been the driving force of the development of the LN from the very beginning : the search term **must** be directly and quickly creatable **from a graphical structure diagram alone, on a personal computer, offline, with no interaction of a mainframe computer necessary**. If we extend this set of conditions from the LN to the creation of chemical nomenclature, then a tool of remarkable power results. The following section deals with progress in this direction, without going into technical details.

Automatic Generation of Systematic Nomenclature

The same user interface which communicates with the LN software can also communicate with a second package, which returns a systematic name in a matter of seconds [4].

One example will suffice:

The structure of Fig.5 was analysed for "NAME". The result returned after a few seconds from the PC was :

6-(8-Bromo-10,11-dihydro-5**H**-dibenzo[**a**,**d**]cyclohepten-10-yl)-5-(3-methyl-but-2-enyloxy)-5,6-dihydro-1**H**-pyrazin-2-one

This name was not simply "found" in a list, but was actually created by the software. This is an important difference. The generation of the name is, like the generation of the LN, completely independent of whether the given structure actually has been reported in the literature or not. It is a general retrieval tool, independent of the database and universally applicable.

Further details of the combined use of nomenclature and LN tools will be published elsewhere in the near future.

AUTONOM 6-(8-Bromo-10,11-dihydro-5#H!-dibenzo[#a!,#d!]cyclohepten-10-yl)-5-(3-methyl-but-2-enyloxy)-5,6-dihydro-1#H!-pyrazin-2-one

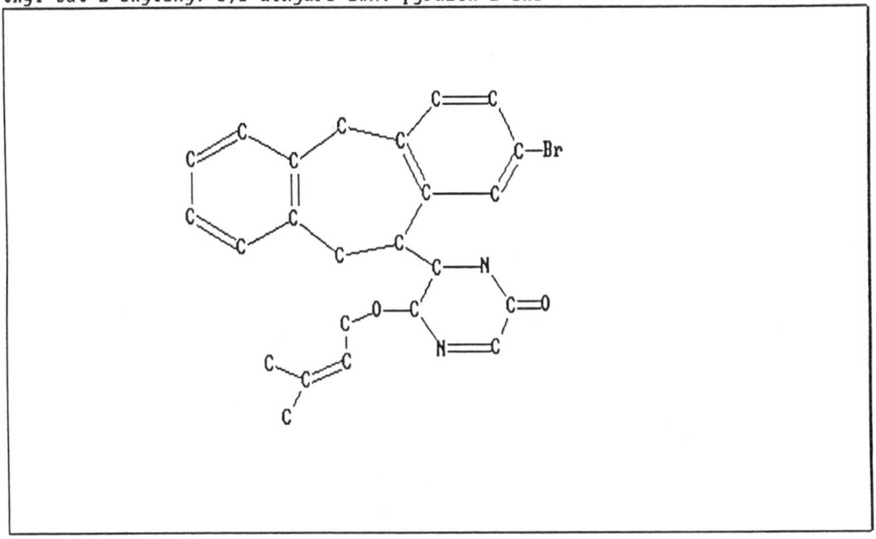

Name Input Exit

Fig. 5

Automatic Generation of Chemical Nomenclature Offline, directl from the structure input on PC-Software

Acknowledgment

The author thanks the German Bundesminister für Forschung und Technologie (BMFT) for generous financial help in the continuation of this research.

Literature cited

1. Lawson, A.J. In Software-Entwicklung in der Chemie 2; Gasteiger, J. Ed.; Springer Verlag: Heidelberg, 1988,p1.

2. Lawson, A.J. In Graphics for Chemical Structures; Warr, W.A. Ed.; ACS Symposium Series 341; American Chemical Society: Washington, D.C., 1987, pp 80-87

3. Santora, N. This Monograph, Chapter 6.

4. Goebels, L., Lawson, A.J. and Wisniewski, J. In Software-Entwicklung in der Chemie 3; Gasteiger, J., Ed. 1989 (In preparation).

RECEIVED May 17, 1990

Author Index

Baker, David C., 130
Baker, Gayle S., 130
Barth, Andreas, 24,113
Haglund, Katharine A., 42
Hartwell, Ieva O., 42
Heller, Stephen R., 1
Jochum, Clemens, 10
Lawson, Alexander J., 143
Ridley, Damon D., 88
Santora, Norman J., 80
Welford, Stephen M., 64

Affiliation Index

Beilstein Institute, 10, 143
Dialog Information Services, 42
SmithKline Beecham Pharmaceuticals, 80
Springer–Verlag, 64
STN International, 24, 113
U.S. Department of Agriculture, 1
University of Alabama, 130
University of Sydney, Australia, 88

Subject Index

A

Abnormal mass lists, BRCT, 19
Academic discount pricing, Beilstein Online searches, 139
Academic use, Beilstein Online, 130–142
Accessing of information, Beilstein database, 21
Acetone cyanohydrin, obtaining reaction information, 92
Acyclic compounds
  Beilstein Online, 131
  main division of Beilstein Handbook, 3*t*
  online availability, 11
Adabas, database management, 20–21
Additional indexes, DIALOG, 46
Adenine, substructure search for derivatives, 32,34*f*
Alkoxyisoquinolines, query structure building using MOLKICK, 70
Alphanumeric fields, factual file, 14
2-Amino-3-iodopyridine, searching for preparation information, 133–135,136*f*
Answer sets
  display using MOLKICK, 76,77–78*f*
Answer sets—*Continued*
  Lawson number searching, 85–86
  structure search on DIALOG, 75–76
Antibiotic activity, pattern recognition, 80–82
ASCII format, ROSDAL strings, 75
Atom code file, structure searching, 68
Atom counts, Lawson number searching, 84–85
Atom list, BRCT, 17
Atom type specification, MOLKICK, 70,72*f*
Attachment point specification, MOLKICK, 70,74*f*
Aziridines, Michael addition, reaction searching on Beilstein Online, 85–86

B

Basic index
  data and information searches, 40
  substance identification, 30,32
  terms, searching complete phrases and words, 46
  text searching, 43,44*t*
Basic Series, Beilstein Handbook, 25

Beilstein
  Fifth Supplementary Handbook Series, Beilstein Online, 12
  Friedrich Konrad
    background, 1
    publication of handbook, 1–2
Beilstein database
  accessing, 21–22
  background, 10
  chemical structure searching on DIALOG using S4/MOLKICK, 64–79
  comparison with CAS registry database, 32,41,65,108–112
  computerizing, 10–22
  contents, 24,143
  features and limitations, 109*f*
  fields containing chemical reaction information, 88,89*f*
  growth in 1990, 41
  implementation on DIALOG, 42–63
  main sources, 10–11
  missing values, 126
  numerical and textual fields, 113–114
  physical property data, 113–129
  searching, 21
  STN implementation, 24–41
  See also Beilstein Online
Beilstein Dictionary, use with Beilstein, 131
Beilstein file cards, contents, 10
Beilstein Handbook of Organic Chemistry
  availability as online database, 1
  contents, 10,42–43
  conversion into computer-readable form, 1
  description, 1–2,25
  history, 1–2
  main divisions, 3*t*
  ordering system, 14
  organization of fourth edition, 2*t*
  reasons for not using, 131
  rules for using, 3
Beilstein Institute
  production of online database, 131
  software development, 11
Beilstein Online
  academic use, 130–142
  advantages, 139,141
  comparison with other factual databases, 135,139,141–142
  data availability, 142
Beilstein Online—*Continued*
  description, 42,131–133
  file crossover to CAS Online, 141
  implementation of database, 4*t*
  limitations, 141–142
  overview 1–8
  positive features, 130
  STN search costs, 139,140*f*
  structure, 132*f*
  support of industry information needs, 80–87
  See also Beilstein database
Beilstein Registry Connection Table
  adoption for inhouse systems, 15
  requirements, 15–16
Beilstein registry number(s)
  assignment to substances, 40
  description, 14
  pyrrole reaction search, 101
  range searching, 90
  resulting from search, 28
  retrieval of factual data, 22
  stereoisomers, 65,66*f*
  substances in Beilstein database, 131
  use in searching Beilstein database, 21
Benzofurans
  nitration, structure searches, 103–107
  sample answer for conversion to furocoumarins, 98*f*
Bibliographic data and information searches, 40
Biological function, inclusion under physiological behavior, 15
Biosynthesis, 32,35,36*f*
Boit, H.-G., editor of Beilstein Handbook, 2
Boolean fields, factual file, 14
BRCT
  adoption for inhouse systems, 15
  criteria governing design, 15–16
Byproducts, inclusion under preparative data, 14

C

Caffeine analogues, search strategies for preparation from pyrimidines, 108–112
Caffeine isolation, Beilstein Online searches, 135,137*f*
Cahn–Ingold–Prelog configuration descriptors, stereospecific searching, 67

Calculation of Lawson numbers, offline, 150
CAOLD, features and limitations, 109*f*
Carbocyclic compounds, main division of Beilstein Handbook, 3*t*
Carbohydrate searching, Beilstein Online, 133
Carboxylic acids, retrieval by Lawson numbers, 145–146
CAS Online
 comparison with factual databass, 139,141
 file crossover to Beilstein Online, 141
 first commerical use, 5*t*
 search costs, 139
CAS registry file
 comparison with Beilstein database, 32,41,65
 online searching, 67
CAS Registry III, tautomerism conventions, 67
CAS registry number(s)
 assignment to substances, 40
 description, 14
 factual databases, 141
 inclusion in fields, 101
 searching for chemical reaction information, 108
 use in searching Beilstein database, 21
CASREACT
 CAS Online, 141
 features and limitations, 109*f*
Catalyst searching, preparation field, 57
Charge lists, BRCT, 18
ChemConnection, use to create ROSDAL strings, 59,62
Chemical(s), requirements for inclusion in Beilstein Handbook, 1
Chemical Abstracts
 complement to Beilstein database, 108–112
 features and limitations, 109*f*
 reaction information, 108–112
Chemical behavior
 description, 15
 reaction scheme, 35,37*f*
Chemical derivative field, Beilstein database, 88
Chemical Information Specialist, use of Beilstein Online, 83–85
Chemical name fields
 basic index, 43,44*t*
 Beilstein database, 90
Chemical name(s)
 indexing, 45
 search field, 28,30
 text searching, 43–46
Chemical reactions
 basic index, 44*t*
 Beilstein database, 32,35–37
 Beilstein Online, 8
 DIALOG, 56–58
 DIALOG vs. STN, 57,58*t*
 fields containing, 56–57
 reaction scheme, 56*f*
 searching for, 88–112
Chemical substance identification, Beilstein database, 30–34
ChemTalk, ROSDAL strings, 59,62
CIP configuration descriptors, stereospecific searching, 67
Citations for Beilstein Handbook, format, 3
Classification, Lawson number analysis, 154
Closed ranges
 numeric data, 116
 restricted searches, 121
CODEN, document retrieval, 135,141
Collected data fields
 basic index, 44*t*
 DIALOG, 50
 use with data present, 50,51
Comment field, basic index, 44*t*
Communication of query structures to DIALOG,
 MOLKICK, 68–69
Compare operators, numeric searches, 117–118*t*
Component analysis, pattern recognition study, 81
Computerizing the Beilstein database, 10–22
Connect time costs, Beilstein Online, 6
Connection tables, use to store structures, 15
Content of database, Beilstein Online, 3–4
Controlled terms
 entering of nonnumeric factual data, 46
 multicomponent system field, property name searching, 28,38,128
 nonnumeric property searches, 50
 physical property searches, 38
 search field, 28
Conversion of units, Beilstein database, 124–126,127*f*

Costs
Beilstein Online on STN, 139,140*f*
database searching, 6–7
Coumarins, sample answer for conversion to furocoumarins, 98*f*

D

DARC
availability, 5
first commerical use, 5
handling of structures and substructures, 15
Data fields
based on molecular formula, 52*t*
basic index, 44*t*
factual databases, 139,141
Data present field
DIALOG, 47
reference parameters, 50*f*
Data present output, DIALOG, 61*f*
Data structure
Beilstein database, 26
defining for Beilstein Online, 11
Database, management by Adabas, 20–21
Database content, Beilstein Online, 3–4
Database design, Beilstein database, 26–30
Database generation, 20*f*
Database implementation, availability of Beilstein Online, 4
Database structure, Beilstein Online, 12–20
Default units, Beilstein database, 125–126
Delocalized charge list, BRCT, 18
Delocalized unpaired valence electron list, BRCT, 19
Derivatives, characterization, 15
DIALOG
availability of Beilstein Online, 4–5
data fields, 44*t*
general discussion, 7
implementation of Beilstein database, 42–63,90
online communication, 68–69
retrieval of tautomers and stereoisomers, 67
search costs, 6
segmentation algorithm, 43,45–46
structure queries, 150
submission of queries, 75
use to search Beilstein database, 22
use with S4 and MOLKICK, 64–79
DIALOG vs. STN, preparations and reactions searching, 57,58*t*
Dictionary term selection, Lawson number searching, 84,86
Dihalo–tellanes, obtaining preparation information, 92
Diketones
search strategies for conversion to pyrroles, 99–102
substructure search, 101
Dipole moment, data structure, 28,30,31*f*
Discounts, Beilstein Online searches, 139
Display command
Beilstein Online searches, 133,135
DIALOG, 76
Messenger software, 126
stereochemical structures, 60,62–63
Display formats
Beilstein database, 26–30
DIALOG, 60–62
Display technique, pattern recognition, 81
Document structure, Beilstein database, 28,29*f*
Drug design, use of Beilstein Online, 80–82

E

Ecological data, inclusion under physiological behavior, 15
Educts, use in searching preparative routes for a compound, 141
Element count field, DIALOG, 52*t*
Energy parameters, subdivision of physical properties, 14
English chemical name field, basic index, 43,44*t*
Episulfides, steroid, search strategies for preparations, 103,104*f*,106*f*
Epoxides, search strategies for conversion to episulfides, 103,104*f*
Ergaenzungswerk fourth edition of Beilstein Handbook, 2
Error messages, MOLKICK, 75
Evaluated data, restricted searches, 35,37*f*,128
Exact value search
restricted, 121,123*f*
schematic, 119*f*
Excerp, use to input structures and factual information, 12

EXPAND command
identification of search terms, 101
listing search terms in text fields, 94
Expand list, property hierarchy, 38,39*f*
Experimental, search term, 121

F

(F) operator, searching collected data fields, 50*f*
Factual databases
Beilstein, 10,12,14,26
comparisons, 135,139,141–142
literature references, 40
STN implementation, 24–41
Field(s), factual databases, 139,141
Field availability
Beilstein Online, 131,133
display format, 26–30
pattern recognition study, 82
physical property searches, 38
preparation of pyridines, 133,135
property name searching, 128
reaction data searching, 32,35
search field, 28
Field not available search, retrieval of missing values, 126,127*f*
File 390 on DIALOG, description, 42
File segmentation, databases, 116
File size, relation to online search time, 68
First implementation phase, Lawson numbers, 144–149
Flavone derivatives, biosynthesis, 35
Flexibility, BRCT, 16
Formula weight, search field, 30
Free display formats, DIALOG, 60
From list, BRCT, 17
Full-word retrieval, comparison with segment retrieval, 45
Full file
Beilstein Online, 3,5
pyrrole reaction search, 101
registered substances, 94
restricted searches, 90
structure, 88,89*f*
Furan
searching for preparations, 95–97,98*f*
searching for reactions, 97,99,100*f*
Furocoumarins
sample answers for conversion of coumarins and benzofurans, 98*f*
searching for preparations, 95–97

G

General fields, data and information searches, 40
Generation of systematic nomenclature, 154,155*f*
Generic groups, MOLKICK, 69–70,73*f*
Generic residues help information, MOLKICK, 70,74*f*
GEOFF
use by DIALOG, 7
use to display structures, 62
German chemical name field, basic index, 43,44*t*
German language, drawback to using Beilstein, 131
Gmelin database
missing values, 126
online availability, 65

H

Handbook, search term, 90,105,121,122*f*
Handbook data, restricted searches, 35,37*f*,128
Hauptwerk\, fourth edition of Beilstein Handbook, 2
Header vector, BRCT, 16–17
Heat of formation specification, physical property searches, 38,39*f*
Heilbron Online, comparison with Beilstein Online, 135,139,141–142
Heterocyclic compounds
Beilstein Online, 131
main division of Beilstein Handbook, 3*t*
online availability, 11
Hierarchical structure, properties in Beilstein database, 114,115*f*
HILIGHT format, DIALOG, 60,62*f*
Hot-Key combination, MOLKICK, 75–76
HTSS
first commerical use, 5*t*
structure searching, 67

Hydrogen isotope lists, BRCT, 19
Hydrogen list, localized, 17

I

Identification of chemical substances, Beilstein database, 30–34
Identification of natural products, Beilstein Online, 82–83
Identifiers, description, 14
Implementation of database, availability of Beilstein Online, 4
Indexing method
  preparations of compounds, 90,91*f*
  reactions of compounds, 92,93*f*
Individual data fields, basic index, 44*t*
Industry information needs, support using Beilstein Online, 80–87
Information forms, Beilstein database, 30
Information levels, Beilstein database, 28,29*f*
Information needs of industry, support using Beilstein Online, 80–87
Information retrieval, Beilstein database, 21
Input of data, Beilstein Online, schedule, 12,13*f*
Input software, Beilstein database, 11
Intermediate file, Beilstein database, 20
Isocyclic compounds
  main division of Beilstein Handbook, 3*t*
  online availability, 11
Isolation chemistry, use of Beilstein Online, 82–83
Isolation from natural product field
  Beilstein database, 88
  caffeine isolation, 135
  natural product identification, 82–83
Isolation of caffeine, Beilstein Online searches, 135,137*f*
Isolation of natural products field, biosynthesis information searching, 35
Isotope distinction, Beilstein Online, 46
IUPAC-oriented nomenclature
  Beilstein Handbook, 25
  Beilstein Online, 30
IUPAC nomenclature rules, Beilstein database, 43

J

Jacobson, P., editor of Beilstein Handbook, 2

K

Keyword(s)
  DIALOG, 46
  physical properties, 120–121
Keyword subfields, use to restrict searches, 38
Knowledge index, availability of Heilbron Online and Merck Index Online, 141–142
Known abnormal mass list, BRCT, 19
KWIC format, DIALOG, 60,62*f*

L

Large information compounds, inclusion in Beilstein Online, 11–12
Lawson number(s)
  average occurrence in Beilstein database, 145*t*
  description, 14,144
  first implementation phase, 144–149
  first versus second implementation, 151
  generation by user at query time, 147
  inclusion in substance records, 54,55*f*
  limitations, 146–147
  molecular formulas, 150
  number of values per compound, 144
  offline calculation, 150
  offline generation and online use, 143–155
  positional isomer searching, 144,149
  range of values, 144
  retrieval tool, 144
  second implementation phase, 149–154
  shape discrimination, 151–154
  use in Beilstein database searches, 32
  use with Beilstein Online, 3
  use with nomenclature, 154
Lawson number creation, SANDRA, 8
Lawson number range
  estimation, 84
  Markush structures, 83–85
  reaction products, 86
Lawson number searching
  Beilstein Online, 83–86
  dictionary term selection, 84,86
  rationale for use, 84

LIC, inclusion in Beilstein Online, 11–12
Licensing, structure file of Beilstein database, 22
Limited search, evaluated data, 35,37*f*
Linkage of terms, substance searches, 94–95
Lists, from connection tables, 16–20
Localized charge list, BRCT, 18
Localized hydrogen list, BRCT, 17
Localized unpaired valence electron list, BRCT, 18
Logical operators, numeric searches, 117–118*t*
Luckenbach, R., editor of Beilstein Handbook, 2

M

MACCS, handling of structures and substructures, 15
Main Series, Beilstein Online, 131
MAP command, use to save Lawson numbers and Lawson number sets, 54,55*f*
Markush searching, Beilstein Online, 83–85
Mass lists, BRCT, 19
Maxwell Online, structure queries, 150
Maxwell Online–ORBIT, availability of Beilstein Online, 4–5
Medicinal chemistry, use of Beilstein Online, 80–82
Merck Index Online, comparison with Beilstein Online, 135,139,141–142
Messenger Search System, Beilstein database, 21
Messenger software
  conversion of units, 38,125–126,127*f*
  creation of structure queries, 133
  numeric data service, 129
  numeric range searching, 117–120
  parameter-dependent searches of properties, 114
  search query for a numeric range, 124
  structure-building commands, 135
Methane, distinguishing between isotopes, 46
5-Methoxyisoquinoline, structure, 70
Michael addition of aziridines, reaction searching on Beilstein Online, 85–86
Missing values, retrieval, 126,127*f*
Molecular descriptor codings, pattern recognition study, 81
Molecular elements field, DIALOG, 52*t*
Molecular formula(s)
  data fields, 52*t*
  search field, 28,30
  use for searching, 21,51–54,101
  use to generate index terms, 32,33*f*
  use with Lawson numbers, 150
Molecular formula field
  DIALOG, 52*t*
  natural product identification, 83
Molecular weight field, DIALOG, 52*t*
MOLKICK
  comparison with other software programs, 69
  creation of structure queries, 133,135,138*f*,141
  DIALOG, 64–65
  generic groups, 69–70,73*f*
  lack of communications software, 69
  query structure editor, 69–78
  ROSDAL strings, 7–8,59,62
  structure drawing functions, 69
  template libraries, 69–70,71*f*
Monoalkoxyisoquinolines, investigation of properties, 70–78

N

Name searching
  Beilstein Online on DIALOG, 43–46
  limitations of Beilstein Online, 141
  STN, 94
Name segments, Lawson number searching, 84–85
Natural product fields, comment and isolation, 43
Natural product identification, use of Beilstein Online, 82–83
Nitration of benzofuran, structure searches, 103–107
NMR spectroscopy, identification of organic compounds, 131
Nomenclature searching
  Beilstein Online on DIALOG, 43–46
  retrieval of positional isomers, 149
  STN, 94
  use with Lawson numbers, 154
Nomenclature segmentation algorithm, DIALOG, 43,45–46
Nondefault valence list, BRCT, 18

Nonevaluated data, restricted searches, 128
Nonnumeric data, searching with reference tags, 49*f*,50
Number of attributes field, DIALOG, 56*t*
Number of chemical reactions field, DIALOG, 56*t*
Number of Lawson numbers field, DIALOG, 56*t*
Number of preparations field, DIALOG, 56*t*
Number of references field, DIALOG, 56*t*
Numeric fields, DIALOG, 46
Numeric property search, example, 38,39*f*
Numeric range entities, numeric data, 116
Numeric range overlap detection, Messenger software, 117–120
Numeric range searching
  physical property information, 117–119
  schematic, 119*f*
  strategies, 117
Numerical fields
  DIALOG, 53*f*,54
  factual file, 12,14

O

Obligatory lists, BRCT, 16–17
Obligatory vector, BRCT, 16
Offline calculation, Lawson numbers, 150
Online communication with DIALOG, 68–69
Online search time, relation to file size, 68
Open-ended range search, schematic, 122*f*
Open ranges
  exclusion from numeric range overlap detection, 121–124
  numeric data, 116
Operators
  (F), searching collected data fields, 50*f*
  (S)
    searching collected data fields, 51*f*
    use in Lawson number searching, 86
    use to restrict searches, 99
Optical activity, pattern recognition study, 82
Optical rotation field
  carbohydrate searches, 133,134*f*
  pattern recognition study, 82
Optional lists, BRCT, 16–20
ORBIT
  availability of Beilstein Online, 4–5
  implementation of Beilstein database, 22
  search costs, 6
Organic chemistry, use of Beilstein Online, 85–86
Output, DIALOG, 60–63
Overlap detection feature, numeric range searching, 117–120
Overload of information, common compounds, 141

P

Partner field, basic index, 44*t*
Pattern recognition
  antibiotic activity, 80–82
  description, 81
PC–PLOT, display of structures from Beilstein Online records, 133
Penicillin G, pattern recognition study, 80–82
Periodic group number field, DIALOG, 52*t*
Periodic index term field, DIALOG, 52*t*
Periodic table row field, DIALOG, 52*t*
Personal computers, communication with DIALOG, 68
Pharmaceutical companies, use of Beilstein Online, 8
Phosphorus pentachloride, obtaining reaction information, 92
Phrase indexed fields, DIALOG, 46
Physical property data
  Beilstein database, 38,39*f*,113–129
  Beilstein Online, 8
  description, 14
  design in Beilstein database, 114*t*
  keyword list, 120*t*
  retrieval, 116
  special features supporting retrieval, 124
Physical property searches, obtaining specific answer sets, 120–121
Physical units, Beilstein database, 124–126,127*f*
Physiological behavior, description, 15
Pi-bonding electron list, BRCT, 17
Piperidines, retrieval by Lawson numbers, 147–149
Pongamia glabra, diketone isolated from, 83
Positional isomer searching
  nomenclature, 149
  use of Lawson numbers, 144,149
Prager, B., editor of Beilstein Handbook, 2

Preparation field
  basic index, 44*t*
  finding preparations of a single substance, 94–95
  finding preparations of related substances, 95,97
  finding reactions of a group of substances, 99–101
  preparation of pyridines, 135
  searching Beilstein database, 88–92,94
Preparation information, display, 90,91*f*
Preparation of pyridines, Beilstein Online searches, 133–135,136*f*
Preparation subfields, biosynthesis information searching, 35
Preparations searching
  compounds, indexing method, 90,91*f*
  DIALOG vs. STN, 57,58*t*
  reaction scheme, 56*f*
  related substances, 95–97
  single substance, 94–95
Preparative data, description, 14
Preparative methods, information retrieval from Beilstein database, 143–144
Preprocessing technique, pattern recognition, 81
Pressure, inclusion under preparative data, 14
Pressure field, DIALOG, 50*t*
Pricing
  Beilstein Online on STN, 139,140*f*
  database searching, 6–7
Primary literature, extraction by Beilstein Institute, 25
Product field, basic index, 44*t*
Properties, included in database records, 26,27*f*
Property hierarchy
  expand list, 38,39*f*
  physical property searches, 38
  property name searching, 128
  search field, 28
Property name searching, 126,128,129*f*
Proximity operation
  chemical name segments, 28,30–32
  chemical names, 45
  numeric searches, 117–118*t*
  parameter-dependent searches of properties, 114
Publication year, file segmentation, 116–117
Pyridine
  factual data included in Beilstein database, 26
Pyridine–*Continued*
  searching for preparation information, 133–135,136*f*
Pyrimidines, search strategies for preparation of caffeine analogues, 108–112
Pyrroles
  search strategies for conversion of diketones, 99–102
  substructure search, 99–101

Q

Quantitative field codes, Beilstein Online, 80
Queries, use of Lawson numbers, 145–154
Query range, numeric range searching, 117
Query structure(s)
  communication to DIALOG, 68–69
  DIALOG, 58,60,75
  saving for future use, 76,78*f*
Query structure editor, MOLKICK, 69–78

R

Radical lists, BRCT, 18–19
Range search
  Beilstein registry numbers, 90
  databases, 116–117
  general structure fragments, 103
  nitration of benzofuran, 105
REACCS database, similarity search mode, 84,87
Reaction field
  Beilstein database, 92–94
  finding reactions of a group of substances, 101,102*f*
  finding reactions of a single substance, 97,99
Reaction information
  Beilstein database, 32,35–37
  Chemical Abstracts\, 108–112
  display, 92,93*f*
Reaction partner, substance preparation, 35
Reaction scheme
  chemical reactions, 35,37*f*,56*f*
  substance preparation, 35,36*f*,56*f*
Reaction searching

Reaction searching
  Beilstein Online, 85–86
  comparison between fields, 97,100*f*
  DIALOG, 56–58
  DIALOG vs. STN, 57,58*t*
  group of substances, 99–102
  limitations, 32
  single substance, 97,99,100*f*
Reactions of compounds, indexing method, 92,93*f*
Reagent, substance preparation, 35
Records
  factual databases, 139
  retrieval, Beilstein Online, 133
Reference parameters
  linkage with a specific property value, 51*f*
  linkage with data present, 50*f*
Reference tags
  DIALOG, 46,50–51
  finding and using to search nonnumeric data, 49*f*,50
Refining the search, substance preparation, 95
Registered substances
  full file searches, 94
  short file searches, 94
Registration process, Chemical Abstracts Service and Beilstein Institute, 40
Registry connection table, requirements, 15–16
Registry numbers, use in range searches, 116
Related terms, DIALOG, 46,48*f*,50
Relational operators, numerical field searching, 53*f*,54
Resonance, treatment in BRCT, 16
Restricted searches
  Beilstein database, 90
  use of keywords, 120–123
Retrieval
  Beilstein database, 143
  Beilstein Online, 133
  missing values, 126,127*f*
  physical property information, 116
Richter, F., editor of Beilstein Handbook, 2
Ring-closure list, BRCT, 17
Ring systems, Lawson number analysis, 151–154
Ring template library, MOLKICK, 70,71*f*
ROSDAL strings
  advantages, 75
  creation, 58–59
ROSDAL strings—*Continued*
  linear-notation structure, 7–8
  MOLKICK, 70,75–76
  sending to a PC, 76
  steps involved in creation, 59*f*
  storage, 75
  use to enter structures as queries into S4 search system, 7–8
Royalty rate, use of Beilstein Online, 6

S

(S) operator
  Lawson number searching, 86
  restriction of searches, 99
  searching collected data fields, 51*f*
S4
  building of search files, 68
  first commerical use, 5*t*
  selective searching of tautomers, 67
  structure searching, 67–68,71*f*
  substructure searching, 58
  use to search Beilstein database, 22
  use with DIALOG, 64–65
Salicyl aldehyde, search for derivatives, 30–32
Salts, characterization, 15
SANDRA
  estimation of Lawson number range, 84
  Lawson number creation, 8
  system number location, 3
SANSS, first commerical use, 5–6
Saving query structures for future use, 76,78*f*
Screen searching, STN, 103
SEARCH command, Messenger software, 126
Search costs, Beilstein Online, 6
Search fields, Beilstein database, 26–30
Search possibilities, Boolean searching, 21
Search query for a numeric range, specification, 124
Search restrictions, Beilstein database, 90
Search software, Beilstein Online, 5–6
Search strategy, substance preparation data, 35,37*f*
Searchable fields for chemical reactions, DIALOG vs. STN, 57,58*t*
Searching
  chemical reaction information, 88–112
  factual data, 43–56
  segment name, Beilstein Online, 43–46

Second implementation phase, Lawson numbers, 149–154
Segment retrieval
Beilstein Online on DIALOG, 43–46
comparison with full-word retrieval, 45
Segmentation algorithm, DIALOG, 43,45–46
Segmentation of files, databases, 116
Segmentation of names, STN, 90
SET command
Messenger software, 126
specification of range of publication years, 116
Shape discrimination, Lawson number analysis, 151–154
Short file
Beilstein Online, 3,5,43
pyrrole reaction search, 101
registered substances, 94
SIC, inclusion in Beilstein Online, 11
Similarity reaction searches, use of Lawson number ranges, 86
Single-word name, Beilstein Online search, 46,47*f*
Single point entities, numeric data, 116
Single value search, schematic, 119*f*
Skeletal shapes, Lawson number analysis, 153
Small information compounds, inclusion in Beilstein Online, 11
Software, data and text searching, 5–6
Software programs
creation of ROSDAL strings, 59,62
structure searching, 68–69
Solvent(s), inclusion under preparative data, 14
Solvent field
basic index, 44*t*
DIALOG, 50*t*
Sources, Beilstein database, 25
Special data fields, DIALOG, 56
Spectroscopic data, Supplementary Series V, 131
Starting materials, inclusion under preparative data, 14
Stereo-atom list, BRCT, 18
Stereo-axis list, BRCT, 18
Stereo-bond list, BRCT, 18
Stereochemical structures
DIALOG, 62–63
display codes, 60,62–63
Stereoisomers, Beilstein registry numbers, 65,66*f*
Stereospecific searching
DIALOG, 22
S4, 67
STN, 21
STN
availability of Beilstein Online, 4–5
conversion of units, 125–126,127*f*
general discussion, 7–8
implementation of Beilstein database, 24–41,90,128–129
implementation of Beilstein Online, 131,139,140*f*
numeric range searching, 117
retrieval of missing values, 126
retrieval of tautomers and stereoisomers, 67
search costs, 6
search of Beilstein database, 21
segmentation of names, 90
structure-building commands, 135,138*f*,141
structure queries, 150
STN Express, creation of structure queries, 105,133
STN vs. DIALOG, preparations and reactions searching, 57,58*t*
Stored range, numeric range searching, 117
Structure(s)
search field, 30
storage as ROSDAL strings, 62
subdivision of physical properties, 14
Structure and substructure searching
DIALOG, 22
STN, 21
Structure file
Beilstein database, 15–20,22,65–67
Beilstein Online, 12–20
STN database, 24–41
Structure moving in the edit window, MOLKICK, 70,72*f*
Structure queries
construction, 133
nitration of benzofuran, 105,106*f*
preparation of caffeine analogues from pyrimidines, 108,110*f*
pyrroles, 99–102
Structure-related data, description, 14
Structure search commands, DIALOG, 60
Structure searching
careful use, 103–107

Structure searching—*Continued*
DIALOG, 58–60,64–79
software programs, 68–69
STN, 103
use of software, 5–6
Subfields
preparation and reaction fields, 99
preparation field searching, 90
reaction field searching, 92–94
Subfile field, DIALOG, 56*t*
Submission of queries to DIALOG, 75
Substance preparation, 35,36,37*f*
Substance searching, STN, 94
Substructure searches
conversions of epoxides to episulfides, 103,104*f*
diketones, 101
finding preparations of related substances, 95
nitration of benzofuran, 105
preparations of episulfides with the steroid nucleus, 103
pyrroles, 99–101
Suffixes, to limit searches, 43,44*t*
Supplementary descriptors list, BRCT, 20
Supplementary Series
Beilstein Handbook, 25
Beilstein Online, 131
fourth edition of Beilstein Handbook, 2
Swiss Online Host Datastar, implementation of Beilstein database, 22
Synonym field, Beilstein database, 90
Systematic nomenclature generation, 154,155

T

T. Purpurea, determining the structure of an ethanol extract, 82–83
Tautomerism
conventions, CAS Registry III, 67
group mobile list, BRCT, 19
information concerning, 14
localized list, BRCT, 19–20
selective searching using S4, 67
treatment in BRCT, 16
Telesystemes–DARC system, use with CAS registry file, 67
Temperature, inclusion under preparative data, 14
Temperature field, DIALOG, 50*t*
Temperature specification, physical property searches, 38,39*f*
Template-feature data matrix, pattern recognition study, 81
Template libraries, MOLKICK, 69–70,71*f*
Text searching, Beilstein Online, 43–56
Tolerances, specification, 124
Total element count field, numeric range searching, 118
Toxicity data, inclusion under physiological behavior, 15
Trial option of the display command, Beilstein Online, 133,135,139,141
Truncation symbols, use in searches, 101
Tryptophane
display of data, 26–28
table of contents, 26,27*f*
Type command, DIALOG, 76

U

Uncertainty, physical properties, 117
Unchecked data, restricted searches, 35,128
Units
physical property data, 38
used in Beilstein database, 124,125*t*,126
Unknown abnormal mass list, BRCT, 19
Unpaired valence electron list, BRCT, 18

V

Variable command
nitration of benzofuran, 105, 106*f*
structure searches, 103–107
Vendors, chemical structure searching software for Beilstein Online, 5*t*

W

Wavelength field, DIALOG, 50*f*

Y

Yields, inclusion under preparative data, 14